High-Speed Photonics Interconnects

Devices, Circuits, and Systems

Series Editor
Krzysztof Iniewski
CMOS Emerging Technologies Inc., Vancouver, British Columbia, Canada

FORTHCOMING TITLES:

High-Speed Photonics Interconnects

EDITED BY

Lukas Chrostowski
Krzysztof Iniewski

CRC Press
Taylor & Francis Group
Boca Raton London New York

CRC Press is an imprint of the
Taylor & Francis Group, an **informa** business

CRC Press
Taylor & Francis Group
6000 Broken Sound Parkway NW, Suite 300
Boca Raton, FL 33487-2742

First issued in paperback 2017

Cover: The figure on the cover is adapted from W. Hofmann, P. Moser, P. Wolf, G. Larisch, W. Unrau and D. Bimberg, 2012, 980-nm VCSELs for optical interconnects at bandwidths beyond 40 Gb/s, Proc. SPIE, vol. 8276, 827605.

Version Date: 20130204

ISBN 13: 978-1-4665-1603-8 (hbk)
ISBN 13: 978-1-138-07159-9 (pbk)

Library of Congress Cataloging-in-Publication Data

High-speed photonics interconnects / editors, Lukas Chrostowski, Krzysztof Iniewski.
   pages cm.— (Devices, circuits, and systems)
   Includes bibliographical references and index.
   ISBN 1-4665-1603-8 (hardback)
   1. Interconnects (Integrated circuit technology). 2. Optical interconnects. 3. Photonics. I. Chrostowski, Lukas.
   TK7874.53.H54 2013
   621.36'5—dc23   2012050914

Visit the Taylor & Francis Web site at
http://www.taylorandfrancis.com

and the CRC Press Web site at
http://www.crcpress.com

Contents

Preface

Dramatic increases in processing power, fueled by a combination of integrated circuit scaling and shifts in computer architectures from single-core to future many-core systems, has rapidly scaled on-chip aggregate bandwidths into the Tb/s range, necessitating a corresponding increase in the amount of data communicated between chips to not limit overall system performance. The two conventional methods to increase interchip communication bandwidth include raising both the per-channel data rate and the I/O number. This book discusses these challenges associated with scaling I/O data rates and current design techniques. It describes the major high-speed components, channel properties, and performance metrics.

Increasing interchip communication bandwidth demand has motivated investigation into using optical interconnect architectures over channel limited electrical counterparts. Optical interconnects with negligible frequency-dependent loss and high bandwidth provide viable alternatives to achieving dramatic power efficiency improvements at per-channel data rates exceeding 10 Gb/s. This has motivated extensive research into optical interconnect technologies suitable for high density integration with CMOS chips. This book details how optical interchip communication links have the potential to fully leverage increased data rates provided through CMOS technology scaling at suitable power efficiency levels.

Lukas Chrostowski
UBC

Kris Iniewski
CMOS ET

Editors

Lukas Chrostowski, Ph.D., is an associate professor in the Electrical and Computer Engineering Department at the University of British Columbia (Vancouver) (www. ece.ubc.ca). Born in Poland, Dr. Chrostowski earned a B.Eng. in electrical engineering from McGill University (Montreal, Canada) and a Ph.D. in electrical engineering and computer science from the University of California at Berkeley. With current research interests in silicon photonics, optoelectronics, high-speed vertical-cavity surface emitting lasers (VCSELs) design, fabrication and testing, optical communication systems, and biophotonics, he has published more than 100 journal and conference publications. Dr. Chrostowski has been serving since 2008 as the codirector of the University of British Columbia AMPEL Nanofabrication Facility. He spent his 2011–2012 sabbatical at the University of Washington, Seattle, with Professor Michael Hochberg and the Institute for Photonic Integration/OpSIS foundry service. Dr. Chrostowski is the Program Director of the NSERC CREATE Silicon Electronic–Photonic Integrated Circuits (Si-EPIC) training program (www. siepic.ubc.ca).

Krzysztof (Kris) Iniewski, Ph.D., is managing R&D at Redlen Technologies Inc., a start-up company in Vancouver, Canada. Redlen's revolutionary production process for advanced semiconductor materials enables a new generation of more accurate, all-digital, radiation-based imaging solutions. Dr. Iniewski is also a President of CMOS Emerging Technologies (www.cmoset.com), an organization of high-tech events covering communications, microsystems, optoelectronics, and sensors. During his career, he has held numerous faculty and management positions at the University of Toronto, University of Alberta, SFU, and PMC-Sierra Inc. He has published over 100 research papers in international journals and conferences. Dr. Iniewski holds 18 international patents granted in the United States, Canada, France, Germany, and Japan. He is a frequently invited speaker and has consulted for multiple organizations internationally. Dr. Iniewski has written and edited several books for Wiley, IEEE Press, CRC Press, McGraw-Hill, Artech House, and Springer. His personal goal is to contribute to healthy living and sustainability through innovative engineering solutions. In his leisure time he can be found hiking, sailing, skiing, or biking in beautiful British Columbia. He can be reached at kris.iniewski@gmail.com.

Contributors

Nicola Andriolli
Institute of Communication, Information
and Perception Technologies (TeCIP)
Scuola Superiore Sant'Anna
Pisa, Italy

Pablo Bianucci
Department of Physics
Concordia University
Montreal, Quebec, Canada

Piero Castoldi
Institute of Communication, Information
and Perception Technologies (TeCIP)
Scuola Superiore Sant'Anna
Pisa, Italy

Isabella Cerutti
Institute of Communication,
Information and Perception
Technologies (TeCIP)
Scuola Superiore Sant'Anna
Pisa, Italy

M. Hadi Tavakoli Dastjerdi
Department of Electrical and
Computer Engineering
McGill University
Montreal, Quebec, Canada

Mehrdad Djavid
Department of Electrical and
Computer Engineering
McGill University
Montreal, Quebec, Canada

Werner H. E. Hofmann
Institute of Solid State Physics and
Center of Nanophotonics
Technical University of Berlin
Berlin, Germany

Ludan Huang
Department of Physics and
Department of Electrical
Engineering
University of Washington
Seattle, Washington

Brian Koch
Aurrion Inc.
Goleta, California

Paul Kohl
Interconnect Focus Center
Georgia Institute of Technology
Atlanta, Georgia

Odile Liboiron-Ladouceur
Department of Computer and
Electrical Engineering
McGill University
Montreal, Quebec, Canada

Lih Y. Lin
Department of Physics and
Department of Electrical Engineering
University of Washington
Seattle, Washington

Zetian Mi
Department of Electrical and
Computer Engineering
McGill University
Montreal, Quebec, Canada

Pier Giorgio Raponi
Institute of Communication,
Information and Perception
Technologies (TeCIP)
Scuola Superiore Sant'Anna
Pisa, Italy

Rajarshi Saha
Interconnect Focus Center
Georgia Institute of Technology
Atlanta, Georgia

Rohit Sharma
Department of Electrical Engineering
Indian Institute of Technology Ropar
Rupnagar, India

Wei Shi
Department of Electrical and
    Computer Engineering
University of British Columbia
Vancouver, British Columbia, Canada

Xu Wang
Department of Electrical and
    Computer Engineering
University of British Columbia
Vancouver, British Columbia, Canada

Jin-Wei Shi
Department of Electrical Engineering
National Central University
Jhongli City, Taiwan

1 Energy-Efficient Photonic Interconnects for Computing Platforms

Odile Liboiron-Ladouceur, Nicola Andriolli, Isabella Cerutti, Piero Castoldi, and Pier Giorgio Raponi

CONTENTS

1.1 INTRODUCTION

Impressive data processing and storage capabilities have been reached through the continuous technological improvement in integrated circuits. Whereas the integration improvement has been closely following Moore's law (doubling of the number of devices per unit of area on a chip every 18 to 24 months), the performance of single computational systems has essentially hit a power wall [28]. To leverage the computational performance, explicit parallelism is exploited at the processor level as well as at the system level to realize high performance computing platforms.

Computing platforms of different types offer tremendous computing and storage capabilities, suitable for scientific and business applications. A notable example is given by supercomputers, the fastest computing platforms, used for running highly calculation-intensive applications in the field of physics, astronomy, mathematics, and life science [25]. Another relevant example is given by data centers and server farms, whose emergence has been driven by the development of the Internet. Data centers not only enable fast retrieval of stored information for users connected to the Internet, but they can also support advanced applications (such as cloud computing) that offer computational and storage services. The increasing quest for information and computational capacity to support such applications is driving the performance growth, which is enabled by the parallelism.

The parallelism allows application tasks to be executed in parallel across multiple distinct processors, leading to a reduction in the execution time and to an increase in the computing platform utilization. To benefit from such advantages, the computing systems should be interconnected through a high-capacity interconnection network. In currently deployed data centers and server farms, the parallelism is achieved by tightly clustering thousands of homogenous servers, either high-end servers or more often commodity computers depending on the computing platforms type [3]. Typically, the numerous racks hosting few tens of servers are connected through a rack switch, in turn connected to a cluster switch, as shown in Figure 1.1, so that each server can communicate with any other server. The communication infrastructure consists of electrical switches typically based on Ethernet (for lower cost and flexibility) or Infiniband protocol (for higher performance). Similarly, in supercomputers, an interconnection network with high throughput and low latency is required for connecting thousands of compute nodes [6]. Recently, the distinction between the two most prominent computing platforms, data centers and supercomputers, has become blurry. Indeed, high-performance scientific computation has been demonstrated in data centers by running tasks in parallel through cloud computing, but the performance of the communication infrastructure is found to lag behind the expectations [45]. Indeed, the performance requisites of high throughput and low latency are stringent especially for the high-performance computations.

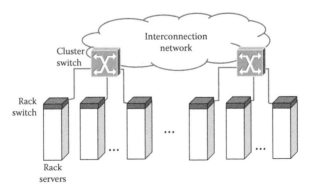

FIGURE 1.1   Generic interconnection network of a high-performance computing platform.

In the last decade, the main bottleneck of computing infrastructure has shifted from the compute nodes to the performance of the communication infrastructure [15]. As computing platforms scale (i.e., increase in the number of servers and in the computational capacity), the requisites of high throughput and low latency are becoming more difficult to achieve and ensure. Indeed, electronic switches are bandwidth limited to the transmission line rate, whereas the number of ports can only scale to a few hundreds per switch. To overcome these limitations, two or more levels of interconnection (i.e., one intrarack, and one or more interrack levels) are required to enable a full connectivity among all the servers. However, the bisectional bandwidth of the interrack level is typically limited to a fraction of the total communication bandwidth. Therefore, to meet the throughput and latency requisites, innovative interconnection solutions that offer a higher degree of scalability in ports and line rate are necessary.

The increase in the size of computing platforms also causes a dramatic increase in power consumption, which may seriously impair further scaling [28]. Currently, the power consumption of large computing platforms is increasing at a yearly rate of about 15% to 20% for data centers [5,7] and up to 50% in supercomputers [26]. According to recent studies [17], the overall power consumed by data centers worldwide has already reached the power consumption level of an entire country such as Argentina or the Netherlands. Within a data center, the communication infrastructure is estimated to drain about 10% of the overall power assuming a full utilization of the servers [1]. However, this assumption is unlikely to occur in today's computing platforms, as the servers are typically underutilized since redundancy is added to ensure good performance in case of failures, especially in computing platforms made out of commodity hardware [3]. When considering recent improvements of the server design to make them more energy proportional (i.e., power consumption proportional to utilization), the network power consumption is expected to reach levels up to 50% of the overall power consumption [1]. Thus, energy-efficient and energy-proportional interconnection solutions are sought.

This chapter explains how photonic technologies can help meet the two crucial challenges of today's interconnection networks for computing platforms, namely, the scalability and energy efficiency. State-of-the-art photonic technologies suitable for connecting and replacing the electronic switches are introduced and discussed in Section 1.2. In particular, interconnection networks realized with photonic devices are proposed for optically switching data between compute nodes or servers of a computing platform. The fundamental strategies for improving the scalability and energy efficiency of optical interconnection networks are presented and explained with examples in Section 1.3. Finally, in Section 1.4, a case study built from recent results obtained in the RODIN* project is used to demonstrate the proposed strategies and the potential benefits of optical interconnection networks for the next-generation computing platforms.

---

* RODIN ("Reti Ottiche Di INterconnessione scalabili ad alta efficienza energetica"—"Réseaux Optiques D'INterconnexions extensibles à haute efficacité énergétique") is a high-relevance bilateral project between Italy and Quebec funded by the Italian Ministry of Foreign Affairs (MAE), Directorate-General for Country Promotion and by the Ministère du Développement Économique, de l'Innovation et de l'Exportation (MDEIE), Programme de Soutien à la Recherche (PSR)–Soutien à des Initiatives Internationales de Recherche et d'Innovation (SIIRI).

1.2 PHOTONIC TECHNOLOGY SOLUTIONS
IN COMPUTING PLATFORMS

To overcome the limitations of electronics, photonic technology solutions for substituting the individual point-to-point links or for replacing the entire electrical switching architectures have been proposed. Photonic communication systems have been shown to achieve large capacity, with low attenuation and crosstalk, and benefit from the data rate transparency of the optical physical layer. Such features are making the photonic point-to-point links excellent replacements of copper cables interconnecting today's electronic switches. Indeed, photonic point-to-point links are already highly used in the new generations of computing platforms [29]. The distance to cover depends on the interconnection level within the system (Figure 1.1). Interconnection of racks over distances up to a few hundred meters can be realized with multimode fiber links. Switching, however, remains in the electronic domain.

To mitigate the current electronic switching limitations, the introduction of optics within the optical interconnection networks has been proposed by the scientific community, and has been shown to achieve greater scalability and throughput compared to electronic switches [14,39]. The design and realization of all-optical interconnection remains, however, challenging due to the lack of effective solutions for all-optical buffering and processing, and due to the nonnegligible power consumption [49].

This section presents the available photonic solutions for both point-to-point links and interconnection networks, the architectural design, and the control strategies for photonic interconnection networks.

1.2.1 Photonic Point-to-Point Links

Compared to electrical links, photonic point-to-point links enable much greater aggregated bandwidth-distance product, allowing increasing communication capacity and reach. Telecommunication systems have made good use of this attribute with aggregated bandwidth-distance in the order of 10^6 Gb/s-m [10,53]. In computing platforms, the number of shared elements and their physical separation distance forced photonic technology development to take a different tangent where bandwidth density requirement becomes as important as the aggregated bandwidth.

A photonic point-to-point consists of the optical source that is modulated by the associated electrical circuitry (e.g., driver), the optical channel, and the photodetector with its associated electrical circuitry (e.g., transimpedance amplifier). For better energy efficiency, photonic point-to-point links not requiring power hungry clock recovery, SerDes, or digital-to-analog converters (DACs) and analog-to-digital converters (ADCs) are preferred. Hence, modulated data rates are often limited to the electrical line rate and are sometimes not resynchronized. Optical sources based on uncooled VCSEL are used for spatial communication with multicore fibers (e.g., ribbons, or multifiber arrays) with one optical carrier per fiber core. Modulated VCSEL (vertical-cavity surface-emitting laser) sources have been shown to meet the electrical line rate of 25 Gb/s [24,29] while being compact, low-cost, and energy efficient. After multimode transmission, *pin* photodetectors convert the signal back to the electrical domain. In recent years, optical active cables (OAC) have been heavily used and have become the prominent solution in the top supercomputers [48]. OACs integrate the optoelectrical (OE) components making them compatible with the electrical sockets at the edge of a server's motherboard and of the electronic switch ports. Futhermore, OAC development enables compatibility with the latest interconnect protocols such as 4x QDR Infiniband [41]. The introduction of optics within the computing platforms is also further pushed inside the servers: the use of optical engines [42] or module-based interconnects eliminates the bandwidth-limited electrical waveguide between the processor and the edge of the board [46]. Indeed, surface-mounted transceivers can be placed adjacent to the processors or other shared elements. Additional improvements in data rates and energy efficiency are possible by monolithically integrating the OE components with the associated electrical circuits on the same substrate [38].

As computing platforms scale, photonic point-to-point links must scale as well, with greater aggregated bandwidth density. As such, wavelength-division multiplexing (WDM) technology must be adopted in the transmission links. Alternatively, multicore fiber development in spatial division multiplexing and multimode modulation in mode division multiplexing have recently attracted great interest in the research community [12]. Complex modulation formats (e.g., OFDM, high order) can also increase the bandwidth, but at the cost of complexity and greater power dissipation due to data processing. Despite the current technological progress, the power consumption of the photonic point-to-point links remains a challenge due to power-hungry signal conversions required at the link terminals [28]. Moreover, the scalability of the overall communication infrastructure with photonic point-to-point links is limited by the processing and switching speed of the electronic switches. To overcome these issues, photonic technologies can be used to switch data in the interconnection networks for more power-efficient computing platforms.

1.2.2 SINGLE-PLANE PHOTONIC INTERCONNECTION NETWORKS

Photonic interconnection networks are proposed as replacement and improvement over today's electronic switches. They provide connectivity between all shared elements of the computing platform (e.g., processors, storage elements) and allow switching in the optical domain. Shared elements can communicate with each other in different ways, either through an allocated time interval to switch data packets (time domain), a designated link (space domain), or an optical bandwidth segment (wavelength domain). By exploiting a single domain for switching, single-plane architectures can be realized. A three-dimensional representation of the domains is shown in Figure 1.2.

1.2.2.1 Space Switching

Optical switching based on space domain exploits the ability to route an optical signal to its destination through one of the multiple distinct data paths available [11]. A dedicated physical path is established to allow concurrent transmission without blocking. The path in the space switch domain is established by properly

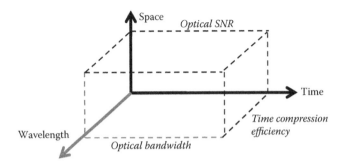

FIGURE 1.2 Three switching domains with their respective scalability limitations. (From O. Liboiron-Ladouceur et al., *IEEE/OSA Journal of Optical Communication and Networking*, vol. 3, no. 8, pp. A1–A11, 2011. With permission.)

setting the photonic devices that act as atomic switches (e.g., Mach-Zehnder–based switches) or gating elements (e.g., semiconductor optical amplifiers, SOA) [51]. The optical power losses experienced by the optical signal when traversing the photonic devices can be compensated through optical amplification. On that front, SOAs are interesting devices to use as they offer both broad gain bandwidth and fast switching time. As amplified spontaneous emission (ASE) noise accumulates, the scalability limitation of the space domain is set by the minimum optical signal-to-noise ratio (OSNR) required by the photodetector for a specified data rate and modulation format. The space domain and its limitation boundary are represented by the *y*-axis in Figure 1.2.

Numerous architectures exploiting space switching have been proposed [11,15]. In single-stage architectures, the optical path traverses a single switching element. A feasible solution is to broadcast the optical signals to multiple destinations, which are then in charge of selecting only the desired signal and discarding the unwanted copies. This type of broadcast-and-select architectures are commonly used for single-plane, single-stage optical interconnection networks [23,35]. In single-plane architectures, the data packets buffered at the input queues (IQ) are transmitted and switched from any of the *N* input ports to any of the *N* output ports. A possible implementation of a 1024 × 1024 broadcast-and-select optical switch based on Spanke architecture is shown in Figure 1.3a and Figure 1.3b and is described in [33]. The binary tree structure of the broadcast-and-select switch exploits the gating characteristics of SOAs. The space-switch can be scaled by increasing the binary tree structure and by adding intermediate amplification stages to maintain the optical signal power after a cascade of power splitters up to the maximum acceptable OSNR degradation. A drawback of this single-stage Spanke architecture is the large number of gating and amplification elements required. To mitigate this issue, multistage architectures (e.g., Benes, Clos) can be exploited where the optical signal is routed through multiple cascaded switching elements reducing the overall number of switching or gating elements [8,51]. Moreover, multistage architectures are often based on small switching elements, typically 2 × 2, simplifying the overall implementation. The main issue with the various proposed multistage architectures is that they require a more complex control for routing all packets [15].

1.2.2.2 Wavelength Switching

The second switching domain, wavelength, exploits the capability of the optical domain to accommodate multiple wavelengths in the electromagnetic spectrum through WDM. To exploit the parallelism nature of the optical domain, the optical bandwidth of the photonic devices (e.g., C-band; 1530–1570 nm) is used to accommodate multiple wavelengths. The wavelength spacing and optical bandwidth of the photonic devices determine the number of channels that can be used to encode the data, which is also strongly dependent on the modulation format and data rate. Greater data rate requiring broader spectra limits the number of carriers over the optical bandwidth. In computing platforms, another limitation is given by the heat and thermal variations. The heat dissipated by the server leads to wavelength drift responses of filters and optical sources. Hence, wider wavelength spacing is required [52]. Wavelength spacing can be reduced when using a complex thermal regulation, at the cost of higher power dissipation. A balance of these adverse effects for an energy-efficient switching could lead to an optimal spacing of 3.2 nm for OOK data rates of 50 Gb/s. With such spacing, the C-band can accommodate up to a few tens of channels, which determine a realistic technological limit of the wavelength domain, as represented by the z-axis in Figure 1.2.

Wavelength-switched architectures taking advantage of WDM technologies are achieved by transmitting the packets on distinct wavelengths, according to the

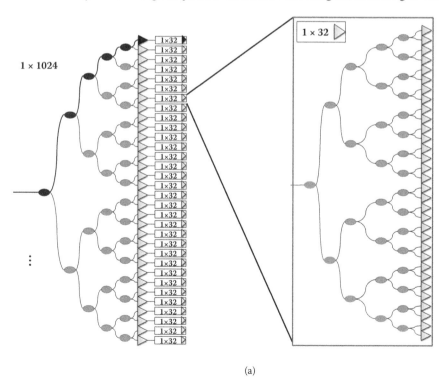

(a)

FIGURE 1.3 (a) 1 × 1024 Broadcast-and-select optical space-switch.

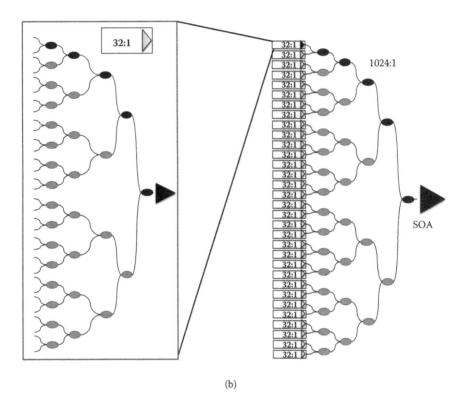

(b)

FIGURE 1.3 (Continued) (b) 1024:1 Broadcast-and-select optical coupler. (From O. Liboiron-Ladouceur et al., *IEEE Journal of Selected Topics in Quantum Electronics*, vol. 17, no. 2, pp. 377–383, 2011. With permission.)

desired destination port. The typical implementation consists of one array of fixed lasers tuned to different wavelengths, an arrayed waveguide grating (AWG), or a coupler that combines the modulated signals coming from the N input ports into a single fiber or waveguide. At destination, another AWG demultiplexes the signal to N output ports with a photodetector for each output port. A limiting factor is given by the wavelength range of the transmitters reducing the maximum number of ports that can be supported to a few tens [19].

1.2.2.3 Time Switching

The third switching domain, time, is exploited by time compressing the packets, then sending them in a time slot allocated for the destination. This dimension is represented by the x-axis in Figure 1.2. In order to guarantee a sustainable throughput, thus avoiding packet dropping at the input buffers, the packet transmission and switching rates must be fast enough to handle all incoming traffic, that is, it should be at least equal to the aggregated rates of the input ports [11]. The scalability of the time domain is further impaired by the overhead necessary for the synchronization. Time switching is typically implemented through a "speed up" of the bit rate, that is, an increase of the transmission rate [16]. Complex modulation formats may be required at the input/output (I/O) off-chip processor interface, making the implementation more difficult and power consuming. An alternative solution consists in the time compression of the packets by exploiting the wavelength domain. Through wavelength striping, serial packets are mapped onto multiple parallel channels (wavelengths) [20,32,43]. By doing so, the transmission time of multiwavelength packets is reduced without requiring a "speed up" of the transmission rate. With this approach, the time domain suffers from the same limitations as the wavelength domain, because the compression in time is determined by the optical bandwidth needed.

In the time-switched single-plane architecture, transmissions of data packets are organized in time frames. Each time frame is divided into a number of time slots equal to the number of output ports, and each output port is identified by a time-slot position within the time frame. One packet is generally sent at each time slot. Data packets are stored in the electronic buffer until a time slot is available for the destination. In a typical implementation of the time-switched architecture, the modulated optical signals coming from the N input ports are combined onto the same fiber or waveguide by a coupler. At the receiving end, the optical signals are broadcast to all the N output ports. A gating element (e.g., an SOA) is inserted before the photodetector at each output port to block the optical signal, except for the desired time slot.

1.2.3 MULTIPLANE PHOTONIC INTERCONNECTION NETWORKS

To overcome the scalability limitations imposed by the use of only one switching domain (Figure 1.2), multiplane architectures can be devised, where multiple switching domains are exploited. Multiplane architectures are organized in a hierarchical and modular way: the interconnection network is composed of multiple cards, each one hosting a number of ports. Output ports in a given card are addressed using one domain while output cards are addressed using another domain, as shown in Figure 1.4a. The modularity of multiplane architectures is extremely useful in computing platforms as servers are typically grouped together in racks as shown in Figure 1.1.

To fully exploit the switching domains in a multiplane architecture, it is necessary to devise architectures and control strategies able to meet the requirements of power consumption [6,40,49,50] and scalability [28] imposed by the current performance growth of computing platforms. Two slotted multiplane architectures have been proposed and studied in the literature [19,33,34,44]. In both, the selection of the output card is based on space switching. The output port is selected using the wavelength or the time domain, leading to a space-wavelength (SW) or space-time (ST) architecture, respectively. Multiplane architectures are able to overcome the limitations of the two separate switching domains and thus benefit of greater scalability than single-plane architectures. Indeed, in the SW architecture, shown in Figure 1.4b, the throughput scalability of the space domain, limited by the minimum required OSNR, is pushed further by the wavelength domain, whereas the time domain benefit is used in the ST architecture as shown in Figure 1.4c. The SW architecture can be implemented in different ways using state-of-the-art photonic devices [19,33,44]. As the physical layer performance is affected by the implementation, the overall

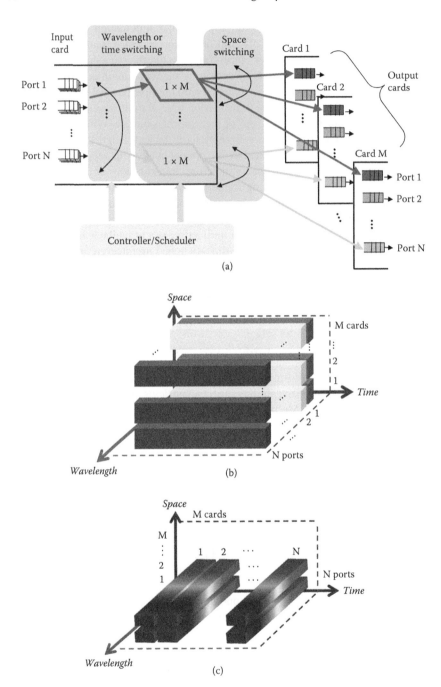

FIGURE 1.4 (a) Multiplane switching architecture. (b) Multiplane switching: Space-wavelength architecture. (c) Multiplane switching: Space-time architecture. (From O. Liboiron-Ladouceur et al., *IEEE/OSA Journal of Optical Communication and Networking,* vol. 3, no. 8, pp. A1–A11, 2011. With permission.)

scalability limitations are also implementation dependent. In the ST architecture, wavelength striping techniques can be used [34] for which the scalability is limited mainly by the physical layer performance of the wavelength striping technique used.

The control of both switching domains in the multiplane architectures is delegated to a scheduler, which takes a scheduling decision at each time slot. Schedulers are also required in single-plane optical interconnection networks but the limited scalability and the simpler implementation makes the scheduling problem easier to solve. In the multiplane architecture, scheduling decisions need to be taken across numerous cards. Moreover, scheduling problems should account for the two different switching domains, which introduce additional constraints to the decision problem. A single centralized scheduler [36] can collect buffer information (e.g., number of stored packets or generation time of the older packets) from each card and take the optimal scheduling decisions while accounting for the domain constraints. However, the centralized scheduler suffers from a scalability problem, as the optimal scheduling decision is computationally intensive and time consuming especially when the number of ports increases. To overcome this issue in multiplane architectures, a distributed synchronous scheduler based on a two-step approach, named two-step scheduler (TSS), has been proposed [44]. TSS addresses the problem of scheduling packet transmission by splitting the problem in two subsequent steps, performed by an intracard and an intercard scheduler, respectively. The intracard schedulers run independently on each card, based on the status of the queues at each input port, without requiring any global information. In particular, each intracard scheduler decides a suitable output port index for any input port on the card. Based on the decisions taken by the intracard scheduler, a single intercard scheduler selects the most suitable output card for each input port. The architecture is properly reconfigured at each time slot, according to the decision taken by a controller. The advantage of the TSS approach is the reduction of the problem complexity as well as a reduction in the latency experienced by the incoming packets in large size networks, with respect to a single-step scheduler. In addition, the TSS approach allows for the parallelization of the scheduling operations, leading to faster computation and higher scalability.

1.3   ENERGY-EFFICIENT PHOTONIC INTERCONNECTION NETWORKS

Photonic interconnection networks are still at their infancy stage of deployment and thus notable innovation opportunities exist in improving the energy efficiency of the interconnection networks and, ultimately, the computing platforms. Different strategies can be pursued in isolation or jointly for improving the energy efficiency of the optical interconnection networks. First, a thorough design of the photonic systems and interconnection architectures along with a careful selection of the constituent photonic devices can achieve significant reduction in the maximum power consumption, while meeting throughput and scalability requirements. Additionally, it has become important to effectively make the power consumption of computing platforms proportional to their utilization, that is, achieve energy proportionality. Therefore, strategies for using the power parsimoniously when operating the photonic interconnects are required. Finally, it is essential that the concepts of energy efficiency and scalability remain coupled to each other when deploying large size interconnects. The different strategies for improving the energy efficiency of photonic interconnection networks are discussed with some practical examples.

1.3.1 ENERGY-EFFICIENT DEVICES

When designing a photonic interconnection network, the first step for improving the energy efficiency is to select the photonic devices that exhibit low power consumption. For this purpose, three important features should be considered: technology availability for device optimization and integration, device thermal sensitivity and cooling requirement, and existence of a low power idle state in active photonic devices.

The technology used for fabricating the photonic devices affects the operating power requirements. However, the electronic devices necessary for driving and controlling the photonic devices are also a major contributor to the power consumption. For example, an optical modulator such as a Mach-Zehnder modulator requires an electrical amplifier with large gain to provide the required voltage swing for the electro-optical effect. Directly modulated optical sources (e.g., VCSEL) or optical amplifiers (e.g., SOA) require an electrical current driver to provide optical gain. Finally, a photodetector includes electrical amplifiers as well as comparators. A recent study actually estimated that the power consumed by the electronic devices can be as high as the power consumed by optical devices in multiplane optical network architectures [9]. But with recent advancements in the CMOS fabrication platform enabling the integration of both the photonic devices and the associated electrical circuitry, the overall power consumption is expected to improve through more efficient device development [2,38]. Also, while active devices such as modulators, amplifiers, and photodetectors are essential in photonic interconnects, passive devices such as filters, couplers, and delay lines should be exploited where possible as they do not consume any energy. Moreover, technology offering the lowest propagation loss should be considered along with design optimization to minimize the insertion loss.

Thermal cooling is often required to limit wavelength drifts and other undesired effects of the physical behavior of the photonic devices especially in WDM-based systems. The removal of heat is an energy-expensive process; it is often estimated that for each watt of generated power, almost another watt is drained for thermal cooling [21]. Therefore, where possible, the selection of uncooled devices is highly recommended for reducing the maximum power consumption, and may lead to tremendous energy savings. Alternatively, systems that are more immune to thermal drift should be favored, hence the necessity to assess the thermal sensitivity of the photonic interconnects performance.

As pioneered in the field of mobile data communication, the exploitation of low-power consumption modes with limited functionalities could significantly reduce the energy usage inefficiencies of computing platforms. Similarly, it would be suitable that the interconnection networks are also enabled with low-power consumption mode. Thus, photonic devices should be designed or enabled for supporting such operating mode(s) with lower power consumption. An example of low-power consumption mode is the idle mode, in which device functionalities are disabled. In idle mode, power consumption should be negligible, but intrinsic power leakage and power drained for keeping the device ready for activation may make idle power appreciable and may still have a significant impact on the overall power dissipation if a large number of idle devices is used [8]. Technology development aiming at minimizing idle state power consumption should therefore be adopted. Moreover, a fast transition from active to idle and vice versa should be ensured to minimize the latency overhead.

1.3.2 ENERGY-EFFICIENT DESIGN OF SYSTEMS AND ARCHITECTURES

The design of the systems and architectures for photonic interconnection networks should be carried out with the goal of ensuring the desired performance with the minimum power consumption. Typically the minimization of the power consumption is achieved by minimizing the number of required optical devices. However, when some photonic devices can be set to idle state or the same architecture can be implemented using different photonic devices, the assessment of the most energy efficient design is not as straightforward.

Energy efficiency of the systems can be further pushed by exploiting passive photonic devices instead of electronics or photonic active devices. For instance, the conventional approach to generate WDM packets uses one modulator–photodetector pair per wavelength. Instead, by exploiting delay lines in a proposed wavelength-striped approach [32], a single modulator–photodetector pair along with their associated circuitry is sufficient for mapping a serial packet onto multiple wavelengths. The power reduction is enabled by using passive photonic devices such as filters and delay lines instead of active devices. However, the passive devices must exhibit low loss propagation to ensure a practical utilization. This is possible with recent technology development based on silicon nitride offering very low loss propagation and low loss bends [4,37]. However, care must be taken to balance the selection of energy-efficient devices with the architecture design and performance. Designing the systems and architectures with only passive devices may lead to high power consumption for amplifiers required to compensate the optical power loss. Therefore, a comprehensive assessment and optimization of the energy efficiency at the system or architectural level is required. For instance, the dual exploitation of the SOA as both an amplifier and gating element has enabled energy-efficient designs [33]. Furthermore, SOAs used as switches have been demonstrated to switch in the subnanosecond range [18,30,31] and can be used in fast reconfigurable broadcast-and-select space switches [23,35]. Although SOAs are power hungry compared to other gating elements, no additional devices are required for compensating the power loss in space switches since they have the amplification capability built-in.

When different designs of the same fundamental interconnection architecture can be derived, the most energy-efficient design can only be found after estimating the power consumption of the alternative designs. An example of architectural design is given by Cerutti et al. [9] where three different implementations of a space-wavelength (SW) multiplane interconnection network are proposed. The implementations differ in the type and also number of exploited devices. The most energy-efficient implementations are those that are less sensitive to negative effects of heating and thus do not require thermal control. Among them, the most energy efficient one is found only after assessing the physical layer performance and accounting for the power drained by the required optical amplifiers.

1.3.3 ENERGY-EFFICIENT USAGE

Energy efficiency should be pursued also through an adequate usage of the photonic interconnection networks, by minimizing the power consumed when transferring and switching data between interconnect ports. First, operating conditions of the photonic devices should be optimized for power dissipation. For instance, in point-to-point photonic interconnects, the optical power generated by the laser can be optimized with the aim of minimizing the power dissipation of the modulator–photodetector pair, as shown in Cho et al. [13]. The optimization accounts for physical impairments, optical power loss, transmission rate, and parameters of the devices, and must ensure adequate physical layer performance.

Another fundamental aspect to consider for an energy efficient usage is the level of utilization of the interconnection network. As computing platforms are underutilized with respect to the computational capabilities to ensure reliability, the capacity of the interconnection infrastructure is also not fully utilized and will fluctuate over time. Without energy-efficient usage, the photonic interconnection network may drain the same amount of power, independently of the utilization level. To avoid a waste of energy, energy-efficient usage should aim at either always operating the photonic interconnection network under full utilization conditions or ensuring that the photonic devices in the interconnect drain power only when utilized. Different approaches to ensure energy proportionality in other contexts have been proposed (e.g., for access networks [27] and electronic interconnection networks [47]). In the specific case of photonic interconnection networks for computing platforms, the adaptability must be instantaneous with very small overhead or performance penalty. As such, the interconnection network should exhibit energy-efficient usage that is agnostic to the utilization and possibly to the scheduler.

Operating the interconnection network at full utilization is rather cumbersome. If the servers (or compute nodes) generate a limited amount of traffic to be switched, the interconnection network can be operated at full utilization only if the traffic of multiple servers is aggregated (disaggregated) at the input (output) ports, for example, multiple servers must be connected to the same input/output port. This can be achieved by a careful planning of the communication infrastructure with the minimum number of interconnection networks, each one operating close to full utilization. Alternatively, a dynamic approach that migrates the processing jobs on fewer servers can be exploited, so that unused servers and interconnects can be switched off [22]. However, some issues are making the exploitation of such techniques difficult. First, the planning of such infrastructure can be complex to realize due to the introduction of an aggregation layer and the operative conditions. Indeed, the planning should ensure that each interconnection network is operating close to full utilization, without exceeding the maximum utilization—a condition

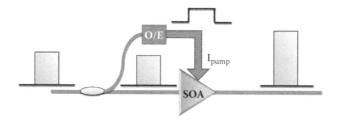

FIGURE 1.5   Energy-proportional SOA gating switch. (From O. Liboiron-Ladouceur et al., *IEEE/OSA Journal of Optical Communication and Networking,* vol. 3, no. 8, pp. A1–A11, 2011. With permission.)

that is difficult to achieve with dynamically changing traffic. Moreover, when approaching the maximum utilization, the waiting time spent by the data packets in the buffers and the related probability of packet dropping may grow beyond the tolerable performance.

It is therefore preferable that the energy efficient usage of the photonic interconnects is achieved at any utilization level by using the minimum amount of energy to transfer and switch data. In other words, the objective is to ensure that the energy consumed is proportional to the amount of switched data, that is, the photonic interconnect is *energy proportional*. Ideally, the power consumption should grow linearly with the utilization and should be null when not utilized. One way to achieve the energy proportionality is by setting to idle state the devices that are not utilized. An example is the proposed SOA-based space switch in [33] where the idle state is self-enabled by the transmitted data or is set by the scheduler (Figure 1.5). The energy-efficient utilization of the SOA with self-enabled techniques has two main benefits. First, the SOA enabling technique operates on a scale of subnanoseconds [30,31]. This permits a quick transition from idle to active state or vice versa with minimal performance penalty. Thus, the space switch closely follows the variations in the traffic load and can rapidly adapt its configuration for better energy proportionality. Moreover, no additional electronic control is necessary for enabling the SOA, thus no additional power for the electronic control is misused.

In practice, the ideal energy proportionality is difficult to achieve due to the nonnegligible power consumption in idle state, power leakages, and difficulties or inability to enable the idle mode in some devices. For instance, the SOA drains a small amount of power when idle. Such amount of power becomes nontrivial for larger interconnects where several SOAs are used in space switches [8]. Another issue is the optical power source, which typically cannot be made energy proportional. Indeed, it is preferable to keep the optical power sources always active as the wavelength stability is not instantaneous and incurs a performance penalty.

1.3.4   ENERGY-EFFICIENT SCALABILITY

The energy efficiency of the photonic interconnection networks should be achieved not only for any utilization level but also for any size of the interconnection network, that is, for any number of ports or cards. In other words, the objective is to achieve a *scalable energy proportionality*: the requirement of energy proportionality should hold for any interconnect size and the consumed energy per bit should stay constant when scaling. Ideally, a scalable energy proportional interconnect consumes the same amount of energy per bit independently of the utilization level and the interconnect size.

Scaling the interconnection network size could in principle even lead to a decrease of the consumed energy per bit. As an example, consider an amplifier that amplifies the optical signals generated by a number of optical sources. When increasing the number of optical sources, the power drained by the optical amplifier can be considered constant while the supported bit rate is increased proportionally to the number of optical sources. As a result, the energy per bit will decrease when scaling the throughput this way. This decrease of the energy per bit is especially advantageous for the photonic devices that are not energy proportional such as lasers.

In practice, scalable energy proportionality is difficult to achieve. The main reason is that the number of photonic devices may not scale linearly with the number of ports. A well-known example is given by the space switches, whose number of switching elements increases more than linearly (even quadratically) with the number of ports [8]. The inherent consequence is that the overall power consumption of the interconnection network increases with the same trend, leading also to an increase of the energy per bit at any utilization level.

To contrast the dramatic growth of the number of devices with the interconnection network size, it is necessary to resort to the energy efficient design techniques discussed in Section 1.3.2. Alternative implementations of the same interconnection architectures may have a better scalability and lead to a more energy proportional scalability.

1.4   RODIN: BEST PRACTICE EXAMPLE

The following section presents two photonic interconnection architectures that make use of the principles and approaches discussed in the previous sections. The two multiplane architectures offer high throughput and scalability, and were developed through the RODIN research project that explores photonic switching technologies for interconnection networks. Both architectures exploit the space domain to address the card, and either the wavelength or the time domain to address the ports (Figure 1.4a). Their energy efficiency is quantified and discussed at the end of the section.

1.4.1   Space-Wavelength (SW) and Space-Time (ST) Switched Architectures

The considered implementation of the space-wavelength (SW) architecture is illustrated in Figure 1.6. Each card is equipped with N fixed lasers, each one transmitting on a different wavelength. Each laser is connected to a modulator, and then directed to a $1 \times M$ optical SOA-based space-switch realized in a tree topology as discussed in Section 1.2.2. On each card, a crosspoint switch with N input queues connects the input ports to the assigned modulators operating on distinct wavelengths. The outputs from the $1 \times M$ space-switches are grouped together according to their destination

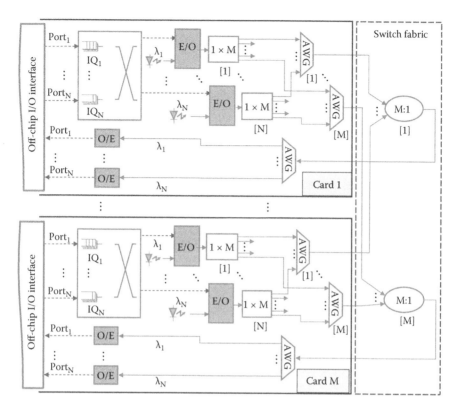

FIGURE 1.6 Space-wavelength (SW) architecture. (From P. Castoldi et al., *Optical Fiber Communication Conference and Exposition (OFC/NFOEC),* 2012; and the *National Fiber Optic Engineers Conference,* pp. 1–3, 4–8, March 2012. With permission.)

card by means of arrayed waveguide gratings (AWG). In the backplane switch fabric, the output of all the AWGs destined to the same output card are coupled together with a M:1 couplers integrated with SOAs to compensate for the losses, as explained in Section 1.2.2. At destination, the signals are demultiplexed by an AWG, and then sent to the optical receivers at the output ports. Each output port on each card is equipped with an optical receiver, so that each output port is uniquely identified in a card by a wavelength. To switch a packet from an input port, the destination card is selected by configuring the $1 \times M$ space-switch, while its destination port on the card is selected by configuring the card's crosspoint so that the packet is transmitted on the wavelength that uniquely identifies the output port. Multiple packets can be simultaneously switched from different input ports to different output ports within the same time slot by properly scheduling the transmissions.

The considered implementation of the space-time (ST) architecture is shown in Figure 1.7. In the ST, the wavelength-striping approach is used to further increase the throughput by encoding packets on multiple wavelengths (also referred to as WDM packets). The ST architecture resorts to the space domain to individually switch WDM packets among cards and the time domain to switch them among different ports of

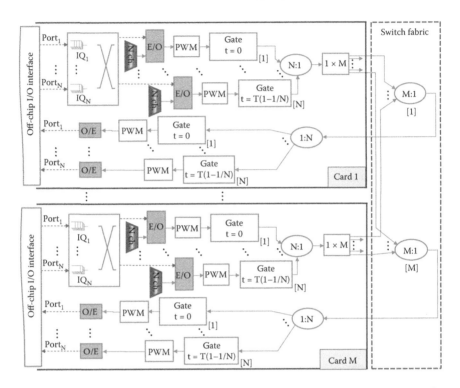

FIGURE 1.7 Space-time (ST) architecture. (From P. Castoldi et al., *Optical Fiber Communication Conference and Exposition (OFC/NFOEC),* 2012; and the *National Fiber Optic Engineers Conference,* pp. 1–3, 4–8, March 2012. With permission.)

a card. The WDM packets are optically generated from the serial electrical packets of duration *T,* as illustrated in Figure 1.8. The bits of the serial packets are used to simultaneously modulate a laser comb of *N* optical channels at the selected data rate. This can be realized with a single broadband modulator that modulates an array of *N* lasers emitting on different wavelengths. Then, a passive wavelength-striped mapping (PWM) element delays each modulated channel by *T/N* from each other and the delayed channels are properly gated in time with an SOA to generate a WDM packet of duration *T/N* as discussed in Section 1.2.3. The serial packet is therefore compressed in time by the number of channels *N,* equal to the number of ports. The WDM packets are transmitted in time slots of duration *T/N* and time multiplexed at the card in a time frame of duration *T,* by an *N*:1 coupler. Each time slot of a time frame is assigned to a specific port of the output card. The multiplexed WDM packets are sent to the SOA-based 1 × *M* switch, which reconfigures itself every *T/N* to direct packets to the proper destination card. A switch fabric as in the SW architecture is used to interconnect cards. The packets destined to a card are grouped by a SOA-based *M*:1 coupler, and then pass through a 1:*N* splitter that broadcasts the routed packets to each port of the card. The SOA gates select the time slot corresponding to the output port followed by the PWM reversely delaying the different channels of the time-compressed WDM packets. Then a single broadband optical receiver converts the now delayed optical signals into a serial packet.

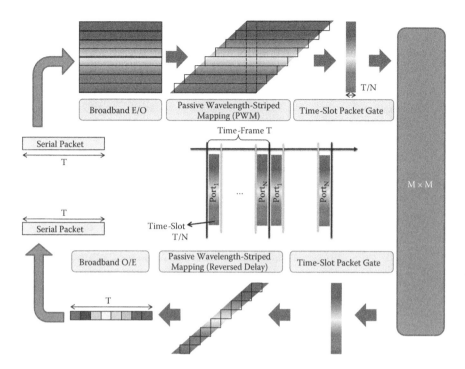

FIGURE 1.8 Wavelength striping in the ST architecture. (From O. Liboiron-Ladouceur et al., *IEEE/OSA Journal of Optical Communication and Networking*, vol. 3, no. 8, pp. A1–A11, 2011. With permission.)

1.4.2 ENERGY-EFFICIENCY ANALYSIS ARCHITECTURE

The energy efficiency of the two multiplane architectures is improved by selecting the most energy-efficient photonic devices that meet the technological requirements of photonic integration. Also, an energy-efficient design has been carried out by exploiting the inherent amplification capabilities of the gating SOAs in the space switches. To enable the energy proportionality, self-enabled SOAs are exploited in the amplification stages of the space switches. When not utilized, the SOAs consume minimum power in idle mode.

The energy efficiency of the two multiplane architectures is compared for line rate $r = 50$ Gb/s. As a reference, a single-plane $M \times M$ SOA-based space-switch interconnection (S) is considered. Before assessing the energy efficiency, the architectures are compared in Table 1.1 in terms of throughput and number of required devices. The multiplane architectures are characterized by a throughput N times higher than the single-plane architecture for the same number of cards. The SW architecture requires N times more $1 \times M$ switches than ST, since a space switch is required for each port in a card. On the other hand, all architectures use the same number of couplers in the backplane switch fabric. The maximum number of ports per card (N) that the ST network can support is constrained by the wavelength-striped technique used for packet time-compression. Since time compression is based on WDM,

TABLE 1.1

Comparison of SW, ST, and S Architectures

	SW	ST	S
Maximum throughput	$MN\cdot r$	$MN\cdot r$	$M\cdot r$
Total ports	MN	MN	M
$1 \times M$ space-switches	MN	M	M
$M{:}1$ couplers	M	M	M
Number of ports per card (N)	8	8	1
Scalability: maximum number of cards at full utilization (<200 pJ/bit)	1024	8192	1024

Source: P. Castoldi et al., *Optical Fiber Communication Conference and Exposition (OFC/NFOEC)*, 2012; and the *National Fiber Optic Engineers Conference*, pp. 1–3, 4–8, March 2012. With permission.

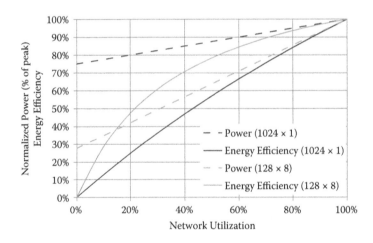

FIGURE 1.9 Normalized power (dashed lines) and energy efficiency (solid lines) versus network utilization of the multiplane (gray lines) and the single-plane (black lines) network architectures.

N is limited by the maximum number of wavelengths that can be used in the optical domain for an adequate physical layer performance. While a large port count is possible, eight ports are chosen in the proposed architecture as no amplification is necessary in the PWM for such number of ports. For a fair comparison, the same number of ports has been utilized for the SW architecture, although this number can easily be scaled to 12 ports.

The normalized power and energy efficiency of the space-switch used in both the ST and SW architectures is shown in Figure 1.9 with respect to the single-plane implementation (1024 × 1), that is, a space-switched single-plane architecture.

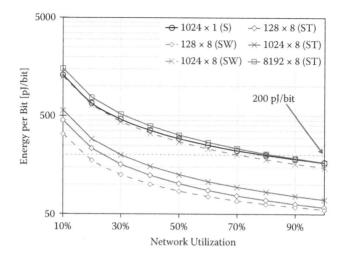

FIGURE 1.10   Energy-per-bit versus network utilization. (From P. Castoldi et al., *Optical Fiber Communication Conference and Exposition (OFC/NFOEC)*, 2012; and the *National Fiber Optic Engineers Conference*, pp. 1–3, 4–8, March 2012. With permission.)

The energy efficiency is obtained by dividing the normalized power by the network utilization. Energy efficiency of 100% corresponds to a complete energy-proportional interconnection network. The figure shows the superior energy-proportionality of multiplane implementations (128 × 8).

The energy per bit is then evaluated for the overall ST and SW architectures, as a function of the network utilization for different sizes $M \times N$ and compared to the S architecture in Figure 1.10. The analysis is achieved using reported power dissipation from recent device implementations [33]. For all architectures, the energy per bit decreases for increasing network utilization (Figure 1.10). Higher power consumption at low utilization is due to the presence of optical devices that are not energy proportional (e.g., lasers), as well as the nonnegligible power consumption of SOAs in idle mode. Among the architectures, the SW architecture is more energy efficient than ST only when the number of cards is low (i.e., $M = 128$ or lower). Power consumption is reduced to approximately one-third for the SW architecture compared to the S architecture for equal throughput (1024 × 1 S compared to 128 × 8 SW). Furthermore, the energy per bit in the ST architecture increases with M more slowly than in the SW. For instance, when increasing the number of cards, M, from 128 to 1024 cards, the energy per bit at full utilization increases by 20% for ST compared to 160% for the SW. The reason for the improved energy scalability of ST is mainly due to the lower number of space switches, as indicated in Table 1.1. In other words, although both architectures rely on space domain for intracard switching, the design of the ST requires N times less switches than in the SW. Thus, unless more energy-efficient designs of the SW architecture are found, the result indicates that ST architecture is not only more scalable in size but also in power consumption. Interestingly, the fully equipped ST architecture (8192 × 8) has power consumption approximately as low as an SW with an 8-fold lower throughput (1024 × 8) and as an S with a 64-fold lower throughput (1024 × 1).

However, from Figure 1.10 it is evident that the energy scalability of the three architectures is still far from the ideal case. As shown in Figure 1.10, in all the considered architectures the energy per bit increases with the number of cards. The main reason is due to the quadratic increase of the number of SOAs in the space switches with the number of cards. Moreover, increasing the number of cards leads to an increase of the optical power losses that need to be compensated by additional amplification stages in the SOA-based switches and couplers, causing a degradation of the power efficiency. For a maximum power consumption set to 200 pJ/bit at full utilization, the SW and S architectures can scale up to 1024 cards, while ST architecture shows the highest scalability, reaching 8192 cards, while still guaranteeing a bit error rate BER less than 10^{-9}. Further improvement of the energy scalability is possible. For instance, alternative architectures of the space switches can be considered [8] or passive switching elements can be used.

1.5 CONCLUSION

Computing platforms require interconnection networks with high degree of scalability and energy efficiency. Optical switching solutions are promising, especially when realizing photonic interconnection networks that exploit multiple switching domains. A new paradigm needs to be instated where careful attention is applied to lower the energy efficiency of photonic interconnection networks. This becomes possible with complementary strategies that involve new design methodologies selecting the most energy-efficient devices, the optimization of the design for energy efficiency, automatic methods for energy-efficient utilization, and solutions for an energy-efficient scalability of the implementation. A best practice example has been presented and shows that the use of multiplane is able to support up to 64 times more ports with respect to the single-plane architecture, with the same energy efficiency.

REFERENCES

1. D. Abts, M. R. Marty, P. M. Wells, P. Klausler, and H. Liu, "Energy proportional data center networks," in *Proceedings of the International Symposium on Computer Architecture (ISCA '10)*, pp. 338–347, 2010.
2. S. Assefa, W. M. J. Green, A. Rylyakov, C. Schow, F. Horst, and Y. A. Vlasov, "Monolithic integration of CMOS and nanophotonic devices for massively parallel optical interconnects in supercomputers," *69th Annual Device Research Conference (DRC)*, pp. 253–256, June 20–22, 2011.
3. L. A. Barroso, and U. Holzle, *The Datacenter as a Computer: An Introduction to the Design of Warehouse-Scale Machines*, Synthesis Series on Computer Architecture, Morgan & Claypool Publishers, May 2009.
4. J. F. Bauters, M. J. R. Heck, D. John, D. Dai, M. C. Tien, J. S. Barton, A. Leinse, R. G. Heideman, D. J. Blumenthal, and J. E. Bowers, "Ultra-low-loss high-aspect-ratio Si3N4 waveguides," *Optics Express*, vol. 19, no. 4, pp. 3163–3174, 2011.
5. C. L. Belady, "In the data center, power and cooling costs more than the IT equipment it supports," *Electronics Cooling*, Feb. 2007.

6. A. Benner, "Cost-Effective optics: Enabling the exascale roadmap," *17th IEEE Symposium on High Performance Interconnects (HOTI)*, pp. 133–137, Aug. 2009.
7. K. G. Brill, "The invisible crisis in the data center: The economic meltdown of Moore's law," Uptime Institute, Technical Report, 2007, white paper.
8. P. Castoldi, P. G. Raponi, N. Andriolli, I. Cerutti, and O. Liboiron-Ladouceur, "Energy-efficient switching in optical interconnection networks," *13th International Conference on Transparent Optical Networks (ICTON)*, pp. 1–4, June 26–30, 2011.
9. I. Cerutti, N. Andriolli, P. G. Raponi, M. Scaffardi, O. Liboiron-Ladouceur, A. Bogoni, and P. Castoldi, "Power and scalability analysis of multi-plane optical interconnection networks," *IET Optoelectronics*, vol. 6, no. 4, pp. 192–200, Aug. 2012.
10. S. Chandrasekhar, A. H. Gnauck, Xiang Liu, P. J. Winzer, Y. Pan, E. C. Burrows, B. Zhu, et al., "WDM/SDM transmission of 10 × 128-Gb/s PDM-QPSK over 2688-km 7-core fiber with a per-fiber net aggregate spectral-efficiency distance product of 40,320 km.b/s/Hz," *37th European Conference and Exhibition on Optical Communication (ECOC)*, pp. 1–3, September 18–22, 2011.
11. H. J. Chao, and B. Liu, *High-Performance Switches and Routers*, Wiley-IEEE Press, May 2007.
12. H. S. Chen, H. P. A. van den Boom, and A. M. J. Koonen, "30 Gbit/s 3 × 3 optical Mode Group Division Multiplexing system with Mode-Selective Spatial Filtering," *Optical Fiber Communication Conference and the National Fiber Optic Engineers Conference (OFC/NFOEC)*, OWB1, pp. 1–3, March 6–10, 2011.
13. H. Cho, P. Kapur, and K. C. Saraswat, "Power comparison between high-speed electrical and optical interconnects for interchip communication," *Journal of Lightwave Technology*, vol. 22, no. 9, pp. 2021–2033, 2004.
14. J. Collet, D. Litaize, J. Van Campenhout, C. Jesshope, M. Desmulliez, H. Thienpont, J. Goodman, and A. Louri, "Architectural approach to the role of optics in monoprocessor and multiprocessor machines," *Applied Optics,* vol. 39, no. 5, pp. 671–682, 2000.
15. W. J. Dally, and B. Towles, *Principles and Practices of Interconnection Networks*, Morgan Kaufmann, 2004.
16. I. Elhanany, and D. Sadot, "DISA: A robust scheduling algorithm for scalable crosspoint-based switch fabrics," *IEEE Journal on Selected Areas in Communications*, vol. 21, no. 4, pp. 535–545, 2003.
17. W. Forrest, J. Kaplan, and N. Kindley, "Data centers: How to cut carbon emissions and costs," McKinsey Report, 2008.
18. C. Gallep, and E. Conforti, "Reduction of semiconductor optical amplifier switching times by preimpulse step-injected current technique," *IEEE Photonic Technology Letters,* vol. 14, no. 7, pp. 902–904, 2002.
19. R. Gaudino, G. Castillo, F. Neri, and J. Finochietto, "Can simple optical switching fabrics scale to terabit per second switch capacities?" *IEEE/OSA Journal of Optical Communication and Networking*, vol. 1, no. 3, pp. B56–B69, 2009.
20. M. Glick, "Optical interconnects in next generation data centers; an end to end view," in *16th IEEE Symposium on High Performance Interconnects (HOTI)* 2008, pp. 178–181, August 2008.
21. A. Greenberg, J. Hamilton, D.A. Maltz, and P. Patel, "The cost of a cloud: Research problems in data center networks," *ACM SIGCOMM Computer Communication Review*, vol. 39, no. 1, 2009.
22. B. Heller, S. Seetharaman, P. Mahadevan, Y. Yiakoumis, P. Sharma, S. Banerjee, and N. McKeown, "ElasticTree: Saving energy in data center networks," in *Proceedings of the 7th USENIX Conference on Networked Systems Design and Implementation (NSDI '10)*.

23. R. Hemenway, R. Grzybowski, C. Minkenberg, and R. Luijten, "Optical-packet-switched interconnect for supercomputer applications," *Journal of Optical Networks*, vol. 3, no. 12, pp. 900–913, 2004.

24. D. G. Kam, M. G. Ritter, T. J. Beukema, J. F. Bulzacchelli, P. K. Pepeljugoski, Y. H. Kwark, Shan Lei, C. W. G. Xiaoxiong, R. A. John, G. Hougham, C. Schuster, R. Rimolo-Donadio, and W. Boping, "Is 25 Gb/s on-board signaling viable?" *IEEE Transactions on Advanced Packaging*, vol. 32, no. 2, pp. 328–344, 2009.

25. S. Kamil, L. Oliker, A. Pinar, and J. Shalf, "Communication requirements and interconnect optimization for high-end scientific applications," *IEEE Transactions on Parallel and Distributed Systems*, vol. 21, no. 2, pp. 188–202, 2010.

26. J. A. Kash, A. F. Benner, F. E. Doany, D. M. Kuchta, B. G. Lee, P. K. Pepeljugoski, L. Schares, C. Schow, and M. Taubenblatt, "Optical interconnects in exascale super-computers," *23rd Annual Meeting of the IEEE Photonics Society 2010*, pp. 483–484, November 7–11, 2010.

27. H. Kimura, N. Iiyama, and T. Yamada, "Hybrid energy saving technique based on operation frequency and active/sleep mode switching in PON system," *23rd Annual Meeting of the IEEE Photonics Society 2010*, pp. 405–406, November 7–11, 2010.

28. P. Kogge, ed., "ExaScale computing study: Technology challenges in achieving exascale systems." DARPA report, September 2008.

29. N. Y. Li, C. L. Schow, D. M. Kuchta, F. E. Doany, B. G. Lee, W. Luo, C. Xie, X. Sun, K. P. Jackson, and C. Lei, "High-performance 850-nm VCSEL and photodetector arrays for 25 Gb/s parallel optical interconnects," *Optical Fiber Communication Conference and the National Fiber Optic Engineers Conference* (OFC/NFOEC), OTuP2, pp. 1–3, March 21–25, 2010.

30. O. Liboiron-Ladouceur, and K. Bergman, "Optimization of a switching node for optical multistage interconnection networks," *IEEE Photonic Technology Letters*, vol. 19, no. 20, pp. 1658–1660, 2007.

31. O. Liboiron-Ladouceur, A. Shacham, B. Small, B. Lee, H. Wang, C. Lai, A. Biberman, and K. Bergman, "The data vortex optical packet switched interconnection network," *Journal of Lightwave Technology*, vol. 26, no. 13, pp. 1777–1789, 2008.

32. O. Liboiron-Ladouceur, H. Wang, A. S. Garg, and K. Bergman, "Low-power, transparent optical network interface for high bandwidth off-chip interconnects," *Optics Express*, vol. 17, no. 8, pp. 6550–6561, 2009.

33. O. Liboiron-Ladouceur, I. Cerutti, P. G. Raponi, N. Andriolli, and P. Castoldi, "Energy-efficient design of a scalable optical multiplane interconnection architecture," *IEEE Journal of Selected Topics in Quantum Electronics*, vol. 17, no. 2, pp. 377–383, 2011.

34. O. Liboiron-Ladouceur, P. G. Raponi, N. Andriolli, I. Cerutti, M. S. Hai, and P. Castoldi, "A scalable space-time multi-plane optical interconnection network using energy efficient enabling technologies [invited]," *IEEE/OSA Journal of Optical Communication and Networking,* vol. 3, no. 8, pp. A1–A11, 2011.

35. T. Lin, K. Williams, R. Penty, I. White, and M. Glick, "Capacity scaling in a multi-host wavelength-striped SOA-based switch fabric," *Journal of Lightwave Technology*, vol. 25, no. 3, pp. 655–663, 2007.

36. N. McKeown, "The iSLIP scheduling algorithm for input-queued switches," *IEEE/ACM Transactions on Networking*, vol. 7, no. 2, pp. 188–201, 1999.

37. A. Melloni, F. Morichetti, R. Costa, G. Cusmai, R. Heideman, R. Mateman, D. Geuzebroek, and A. Borreman, "TriPleX™: A new concept in optical waveguiding," *Proceedings of the 13th European Conference on Integrated Optics*, pp. 3–6, 2007.

38. A. Mekis, S. Gloeckner, G. Masini, A. Narasimha, T. Pinguet, S. Sahni, and P. De Dobbelaere, "A grating-coupler-enabled CMOS photonics platform," *IEEE Journal of Selected Topics in Quantum Electronics*, vol. 17, no. 3, pp. 597–608, 2011.

39. D. A. B. Miller, "Rationale and challenges for optical interconnects to electronic chips," *Proceedings of the IEEE*, vol. 88, no. 6, pp. 728–749, 2000.

40. D. A. B. Miller, "Device requirements for optical interconnects to silicon chips," *Proceedings of the IEEE*, vol. 97, no. 7, pp. 1166–1185, 2009.

41. A. Narasimha, S. Abdalla, C. Bradbury, A. Clark, J. Clymore, J. Coyne, A. Dahl, et al., "An ultra low power CMOS photonics technology platform for H/S optoelectronic transceivers at less than $1 per Gbps," *Optical Fiber Communication Conference and the National Fiber Optic Engineers Conference (OFC/NFOEC)*, OMV4, pp. 1–3, March 21–25, 2010.

42. H. Nasu, "Short-reach optical interconnects employing high-density parallel-optical modules," *IEEE Journal of Selected Topics in Quantum Electronics*, vol. 16, no. 5, pp. 1337–1346, 2010.

43. H. Onaka, Y. Hiroshi, K. Sone, G. Nakagawa, Y. Kai, S. Yoshida, Y. Takita, K. Morito, S. Tanaka, and S. Kinoshita, "WDM optical packet interconnection using multi-gate SOA switch architecture for peta-flops ultra-high-performance computing systems," *Proceedings of the European Conference on Optical Communication (ECOC)*, pp. 57–58, September 24–28, 2006.

44. P. G. Raponi, N. Andriolli, I. Cerutti, and P. Castoldi, "Two-step scheduling framework for space–wavelength modular optical interconnection networks," *IET Communications*, vol. 4, no. 18, pp. 2155–2165, 2010.

45. J. J. Rehr, F. D. Vila, J. P. Gardner, L. Svec, and M. Prange, "Scientific computing in the cloud," *Computing in Science and Engineering*, vol. 12, no. 3, pp. 34–43, 2010.

46. L. Schares, D. M. Kuchta, and A. F. Benner, "Optics in future data center networks," *18th IEEE Symposium on High Performance Interconnects (HOTI 2010)*, pp. 104–108, 2010.

47. V. Soteriou, and L. S. Peh, "Exploring the design space of self-regulating power-aware on/off interconnection networks," *IEEE Transactions on Parallel and Distributed Systems*, vol. 18, no. 3, pp. 393–408, 2007.

48. Top 500 supercomputer sites, www.top500.org.

49. R. S. Tucker, "The role of optics and electronics in high-capacity routers," *Journal of Lightwave Technology*, vol. 24, no. 12, pp. 4655–4673, 2006.

50. R. S. Tucker, "Green optical communications—Part II: Energy limitations in networks," *IEEE Journal of Selected Topics in Quantum Electronics*, vol. 17, no. 2, pp. 261–274, 2011.

51. A. Wonfor, H. Wang, R. V. Penty, and I. H. White, "Large port count high-speed optical switch fabric for use within datacenters [invited]," *IEEE/OSA Journal of Optical Communications and Networking*, vol. 3, no. 8, pp. A32–A39, 2011.

52. F. Xia, M. Rooks, L. Sekaric, and Y. Vlasov, "Ultra-compact high-order ring resonator filters using submicron silicon photonic wires for on-chip optical interconnects," *Optics Express*, vol. 15, no. 19, pp. 11934–11941, 2007.

53. X. Zhou, L. Nelson, P. Magill, R. Isaac, B. Chu, D. W. Peckham, P. Borel, and K. Carlson, "8x450-Gb/s, 50-GHz-Spaced, PDM-32QAM transmission over 400 km and one 50 GHz-grid ROADM," *Optical Fiber Communication Conference and the National Fiber Optic Engineers Conference (OFC/NFOEC)*, March 2011.

2 Low-Loss, High-Performance Chip-to-Chip Electrical Connectivity Using Air-Clad Copper Interconnects

Rohit Sharma, Rajarshi Saha, and Paul A. Kohl

CONTENTS

2.1 INTRODUCTION

Modern electronic systems are composed of a dense fabric of interconnected lines that connect the electronic devices on a chip and chips on boards. The scaling of transistors to small size has resulted in miniaturization of individual devices and their interconnects. Smaller cross-sectional areas of electrical interconnects result in higher signal attenuation, and makes data and clock recovery complex. High-speed interconnects have two primary loss mechanisms: metallic skin effect losses and dielectric losses. Skin effect loss is a phenomenon where, at high frequencies, the current is crowded into the near-surface region of the metal around the periphery of the conductor. As a result, the effective alternating current (AC) resistance per unit length of conductor increases with frequency. The second loss mechanism deals with dielectric losses. Dielectric losses are directly proportional to frequency and are therefore more pronounced at higher frequencies. Also, resistive losses in the return ground path cannot be ignored and are determined by the geometry and materials constituting the return path [1]. Overall, dielectric losses dominate at frequencies above several gigahertz, as shown in Figure 2.1. This figure was produced with data from the International Technology Roadmap for Semiconductors (ITRS) for signal frequencies up to 50 GHz. As shown in Figure 2.1, the dielectric loss is more than twice that of the conductor loss at 20 GHz and the dielectric loss becomes more dominant at frequencies above 20 GHz. Thus, in order to achieve meaningful reduction in signal attenuation it is important to find ways to mitigate the dielectric loss at high frequencies. One way of achieving this goal is to use better interconnect and substrate materials.

Historically, interconnect technology used Al and SiO_2 as the conductor–substrate pair. The transition from aluminum to copper as the interconnect conductor has been one of the most important technological advances in interconnect technology. The fundamental problem with Al-SiO_2 interconnect includes the higher resistance and capacitance, and nonplanarity compared to copper and low dielectric constant interconnect (Cu-low k) technologies. The resistance-capacitance (RC) product represents the characteristic time constant of a series resistor–capacitor circuit. Typically, the bulk resistivity of Cu is less than that of Al, as shown in Table 2.1, although the actual resistivity of the Cu used is somewhat higher than the theoretical limit. This implementation of both Cu and low-k dielectrics results in up to 50% decrease in RC wiring delay compared to Al-SiO_2 interconnect, as shown in Figure 2.2. Since Cu has lower resistivity the effect of line inductance is even more significant in Cu interconnect lines. At higher frequency, the inductive effect dominates causing signal overshoots and ringing and reflections due to impedance mismatch. Complex responses such as these generally require modeling of self and mutual inductance terms. Also, copper has 10 to 100 times higher electromigration resistance compared to aluminum making it a more reliable interconnect material.

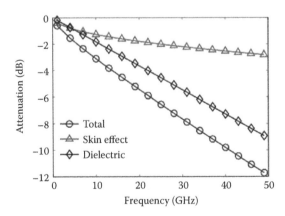

FIGURE 2.1 Conductor and dielectric loss as a function of frequency. (© 2011 IEEE. Reprinted with permission.)

TABLE 2.1
Resistivity of Different Metals

Metal	Bulk Resistivity ($\mu\Omega$-cm)
Silver	1.63
Copper	1.67
Gold	2.35
Aluminum	2.67
Tungsten	5.65

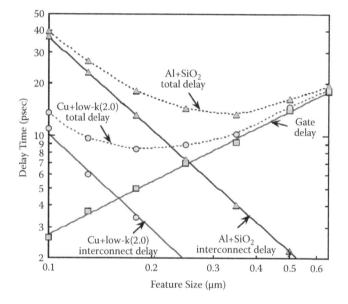

FIGURE 2.2 Delay comparison between interconnects based on copper-low-k and aluminum-SiO$_2$. (© 1995 IEEE. Reprinted with permission.)

Figure 2.2 provides several important insights into the importance of interconnect scaling. One important design metric is the total delay (interconnect and gate delay). Transistor scaling improves transistor delay but not interconnect. The benefits of interconnect with lower RC delay becomes more important with time as interconnect becomes the dominant contributor to system delay with time. This forms the basic motivation for the development and use of low-k materials. Some of the physical attributes of low-k material are listed below:

- Electrical—Relative permittivity (ε_r) <3, near-zero loss, and isotropic
- Mechanical—Excellent adhesion with metals and dielectrics
- Thermal—Low thermal expansion, ideally matched to copper, and high thermal stability

- Chemical—Near zero water uptake, resistant to oxidation (especially oxygen), and does not induce corrosion in metals
- Commercial—Environmentally safe and cost effective

The list of low-k dielectrics include inorganic (glass), organic (polyimide, Teflon), hybrid (organic–inorganic blends), and porous materials made from organic or inorganic insulators [2]. Using vacuum, gas, or air as the dielectric offers an interesting, ultralow-loss option. However, in order to achieve mechanical stability, one can only employ the partial use of an air-clad interconnect structure because of the need for mechanical integrity.

2.1.1 AIR-GAP STRUCTURES FOR REDUCTION OF LINE CAPACITANCE

Previous reports have explored the role of air-gap structures in lowering line capacitance [3–9]. When depositing SiO_2, air gaps or voids can be formed between metal lines. A basic air-gap structure has been presented by Shieh et al. [3]. The interlayer SiO_2 provided structural and thermal stability for the interconnect stack. Unlike aerogels, fluorinated SiO_2, polymers, and other low-k dielectrics, air-gap structures required no additional etching or chemical mechanical polishing. This led to simpler process integration. Simulation results show that reduction in capacitance realized by use of an air-gap structure is comparable to the capacitance reduction by use of homogeneous low-k materials. The use of air gaps reduces the effective dielectric constant (ε_{eff}) [5]. This can be explained by referring to the following equation:

$$\varepsilon_{eff} = \frac{\varepsilon_r \left(h_a + h \right)}{h + \varepsilon_r h_a} \tag{2.1}$$

Here ε_r is the permittivity of the substrate, while h and h_a are the heights of the substrate and air cavity, respectively. Thus, the effective permittivity may approach unity as the air cavity is maximized. One can expect similar results in the case of on-chip, air-gap interconnect as well. Park et al. [6] report nearly 40% reduction in effective dielectric constant by using air-gap structures. If the intralayer air gaps (the dielectric material between signal lines on one layer) are made to be extended into the interlayer (the dielectric layer between metal layers), the reduction in capacitance can be made to be even greater. For a multilevel interconnect structure, intralayer capacitance reduction of approximately 40% can be achieved by using air-gap structures [7]. If the air gaps and dielectrics are deposited effectively, then significant capacitance reduction can be achieved without compromising on the mechanical stability of the structure. One way of doing this would be to deposit air gaps between narrow lines where the maximum capacitive coupling occurs while depositing the conventional dielectric between lines with wider spacing. Extending the air cavity below the plane defined by the metallic structure has a twofold advantage. It can result in lowering the line capacitance and allow the use of wider lines, thereby reducing the conductor losses, as will be discussed later in this chapter. This is particularly important in chip-to-chip interconnects that may transmit signals at frequencies in excess of 50 GHz.

2.1.2 Fabrication of Air-Gap Structures

The fabrication of the air-gap structure is a challenging aspect of this technology. Several processing techniques have been disclosed for forming air isolation in electronic devices including the use of chemical mechanical polishing (CMP) and via etch [4], etchback technique [7,8], sacrificial polymers [5,6,10–13], and wet etching [9]. Among these techniques, the use of a sacrificial polymer for creation of air gaps is a promising method that can be applied to a variety of applications. The fabrication of air-gaps in electrical interconnects based on a sacrificial polymer and plasma-enhanced chemical vapor deposited (PECVD) SiO_2 has been presented by Kohl et al. [10,11]. A sacrificial polymer embedded between metal interconnect structures was used. The air cavity was fully encapsulated by an overcoat dielectric. When thermally decomposed at a controlled rate, the gaseous products of the sacrificial polymer diffused through the dielectric coating thus forming buried air cavities with negligible residue. This technique is well suited for formation of air gaps with sizes ranging from nanometer dimensions to centimeter dimensions and has important applications in the fabrication of air-clad interconnects. The process flow for fabrication of air-gap structures using a sacrificial polymer was presented by Spencer et al. [5] and is shown in Figure 2.3. The thermal decomposition of the sacrificial polymer takes place at a modest temperature (<200°C), which makes it a suitable candidate for chip-to-chip interconnects over organic substrates, such as FR4.

The reliability issues of air-gap interconnects need to be carefully evaluated. These include thermal reliability [4], electromigration reliability [14,15], dielectric reliability [16], process-induced stresses [17], moisture uptake, and corrosion of metal wiring [18]. Joule heating in the interconnect stack also needs to be addressed.

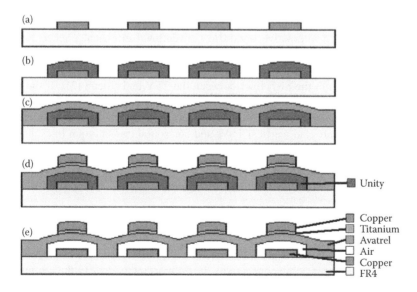

FIGURE 2.3 Process flow diagram of the buildup procedure for an air-gap structure. (© 2007 IEEE. Reprinted with permission.)

The temperature rise in air-gap interconnect is comparable to that in homogeneous SiO_2-based interconnect. It can be seen that the thermal performance of interconnect stacks with air gaps is far superior to any homogeneous low-k material. One would expect a higher metal excursion in air-gap-based interconnects due to a very thin sidewall passivation of SiO_2. However, experimental results show longer electromigration lifetimes for air-gap structures [14]. In summary, air-gap interconnects exhibit comparable or better thermal, mechanical, and electromigration reliability properties when compared to homogeneous SiO_2-based interconnects. Interested readers are referred to the available literature for further insight into this topic.

2.1.3 Design Issues in Chip-to-Chip Interconnects

Chip-to-chip interconnects are planar transmission line structures with lengths several times greater than on-chip interconnects. Thus, the primary factor affecting signal integrity is the signal loss in the interconnect line. As explained earlier in this chapter, there are two major loss mechanisms that occur in chip-to-chip interconnects: the conductor loss due to skin effect and dielectric loss due to lossy substrates. Since the dielectric loss is a stronger function of frequency, it tends to dominate the overall loss mechanism at higher frequencies [19–21]. In the case of chip-to-chip interconnects, bandwidth density and energy/bit are important design metrics to ensure optimum signal integrity. Kumar et al. [20] proposed analytical models that can be used to establish performance limits for air-clad transmission line interconnects for chip-to-chip interconnects. About 5 times improvement in bandwidth density and energy/bit were shown to be feasible by using air instead of GETEK as a dielectric material. From the point of view of interchip connectivity, this is encouraging.

Recent advances in interposer technology have necessitated the use of through silicon vias (TSVs) that form an integral part of 3D interconnect chip-to-chip pathway. TSVs offer an alternative option to conventional electrical connections, as found in interposers and substrates. At higher frequency, as in the case of planar transmission lines, copper-filled TSVs suffer from losses due to skin effect and lossy substrate effects from the SiO_2 liner between the TSV and silicon [21]. If the lossy SiO_2 liner in these TSVs is replaced by air, one can expect performance improvements similar to those found with other air-clad applications. However, the design and fabrication of air-clad, vertical TSVs and horizontal transmission lines remains a research and development topic. The next section highlights important developments in chip-to-chip interconnects for 2D and 3D systems.

2.2 PERFORMANCE IMPROVEMENT USING AIR-CLAD CHIP-TO-CHIP INTERCONNECTS

The basic structure of the air-gap interconnect will be presented before analyzing the improvement in chip-to-chip connectivity. It is important to identify the key metrics that determine the performance of these interconnects, as discussed in this section. This is followed by discussion on performance improvements realized using air-clad interconnects reported in the literature.

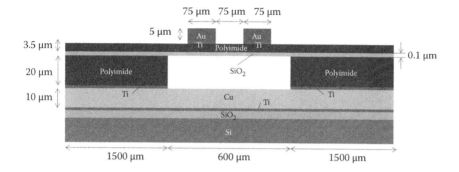

FIGURE 2.4   Schematic of the proposed air-clad differential transmission lines.

2.2.1   Basic Schematic of Air-Clad Interconnect

A fully air-clad interconnect structure would have the conventional, solid dielectric structure completely replaced by the air cavity. However, a hybrid design is more appropriate for mechanical stability where the interconnect lines are placed over a thin membrane of dielectric above an air cavity. Such a design will provide the needed mechanical support required to support the metal lines [21]. The schematic of a suspended air-clad differential pair is shown in Figure 2.4. The dimensions used by Sharma et al. [21] were selected so that the interconnect has a differential impedance of 100Ω. The metal lines were deposited over a thin membrane of polyimide and SiO_2. The supporting polyimide columns along with this membrane create an air cavity that is several times the line width. As discussed later in this chapter, this hybrid design significantly lowers the dielectric and conductor losses. The fabrication of such an interconnect geometry will be briefly discussed in Section 2.3. Furthermore, chip-to-chip interconnect over an Si interposer in an atypical 3D design requires the use of both planar and vertical links, as discussed in Section 2.3. High frequency equivalent circuit and loss models for TSVs use bulk silicon and oxide layers surrounding the TSVs, which contribute to the overall losses in TSVs. The design goal is to create an equally low-loss TSV matching that of the air-clad transmission lines. The proposed schematic of the air-clad TSV structure along with typical dimensions as suggested by Sharma et al. [21] is shown in Figure 2.5. At high frequencies, similar to planar transmission lines, copper-filled TSVs suffer from losses due to skin effect and a lossy substrate because of the use of an SiO_2 layer between the TSV and silicon. Replacing the SiO_2 liner with air can help reduce the losses due to the adjoining substrate. The copper pillars that form the conducting part of the via are held firmly by an SiO_2 pad at the bottom. The top pad is formed from electroplated copper. The process flow and a novel fabrication process for this air-clad TSV is presented in Section 2.3.

2.2.2   Design and Optimization of Low-Loss High-Speed Links

The two main loss mechanisms in chip-to-chip interconnects are the dielectric and the conductor losses. The conductor loss is due to skin effect that results in nonuniform charge distribution resulting in current flow at the surface of the conductor.

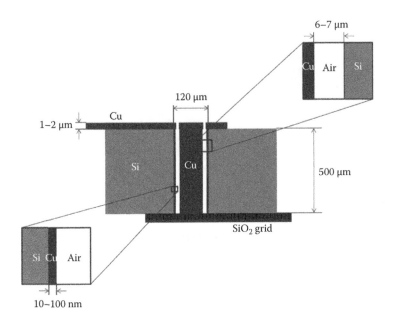

FIGURE 2.5   Schematic of the proposed air-clad TSV.

The characteristic dimension of the skin depth in the metal conductor is the value where the current is reduced to $1/e$ or 37%, as given by

$$\delta = \sqrt{\frac{2\rho}{\omega\mu}} \qquad (2.2)$$

where, δ is the skin depth, ρ is the resistivity of the conductor, ω (or $2\pi f$) defines the frequency (f), and μ is the permeability. Thus, the conductor loss is inversely proportional to the square root of frequency. On the other hand, the dielectric loss is governed by the loss tangent as given by

$$\alpha_{\text{dielectric}} = \frac{4.34}{c}\omega \cdot \tan\delta\sqrt{\varepsilon_r} \qquad (2.3)$$

Here c is the speed of light in free space and ε_r is the relative permittivity of the material. The losses due to the dielectric material are directly proportional to the frequency. Therefore, dielectric losses dominate the overall loss mechanism in chip-to-chip interconnect at gigahertz frequencies, as shown in Figure 2.1. As per the ITRS projection, off-chip signal frequencies may increase to 50 GHz and beyond in the near future. The concern is that the signal integrity may be severely degraded at high data rates for a given energy budget at these frequencies [19–20].

The performance of chip-to-chip interconnects can be evaluated in terms of bandwidth density and energy/bit, as proposed by Kumar et al. [20]. They have reported an analytical model for computing the bandwidth density and energy/bit in an air-clad differential pair with simplistic input/output (I/O) circuits for the transmitter and receiver. The power dissipation models were then used to optimize the bandwidth density and energy/bit simultaneously. Further, a compound metric of bandwidth-density/energy-per-bit was used to maximize the bandwidth density and minimize the energy consumed. This is an important metric that can be used to optimize the interconnect geometry. Figure 2.6 gives this compound metric as a function of interconnect length and width. Traditional substrates based on fiberglass epoxy (i.e., FR4, BT resin, and GETEK) can be lossy. Compared to these materials, the use of air as dielectric material results in significant improvements in bandwidth density and energy/bit, which is shown in Figure 2.7. One can see about five times improvement over standard GETEK material. However, the numbers reported here are for an ideal case. A more realistic structure is shown in Figure 2.4, wherein the improvements may be slightly less due to the presence of a lossy membrane of polyimide and SiO_2. Further, the conductor loss in air-clad interconnects is also reduced due the increased width of the lines. This is done to maintain the same differential impedance for both air and GETEK cases. The HFSS results corroborate this observation. However, the available models do not account for losses in the return path.

A TSV can be seen as a microstrip line on an Si substrate separated by an SiO_2 liner. It supports three fundamental modes of propagation, namely, slow-wave, quasi-TEM, and skin-effect modes [22]. The slow-wave mode can exist at moderate frequency and conductivity in the Si substrate. This mode occurs due to the strong interfacial polarization across the SiO_2 liner resulting in a velocity of propagation much slower than the silicon substrate due to a phenomenon called the Maxwell-Wagner effect. The Maxwell-Wagner effect increases the effective permittivity at lower frequencies. For a typical TSV with a frequency response from a few hertz to several gigahertz, the slow-wave effect is very important for achieving

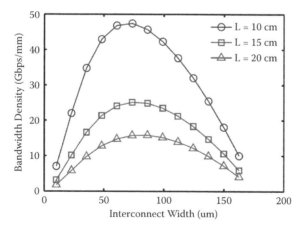

FIGURE 2.6 Maximum bandwidth density and minimum energy/bit. (© 2011 IEEE. Reprinted with permission.)

signal integrity. We can establish the transition from slow-wave mode to quasi-TEM mode by plotting the loss tangent as a function of frequency. One can see the loss tangent going very high at a moderate frequency (typically 1 GHz) where the slow-wave mode transitions into a diffusion type TEM mode followed by a quasi-TEM mode, thus resulting in severe signal attenuation. Air-clad TSVs can help reduce this attenuation, as shown in Figure 2.8.

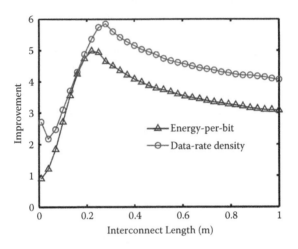

FIGURE 2.7   Improvements in key design metrics for air-clad interconnects. (© 2011 IEEE. Reprinted with permission.)

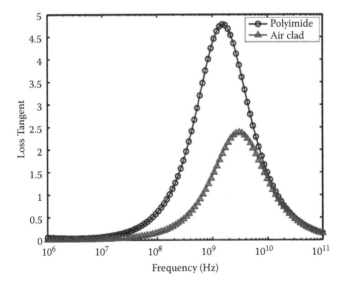

FIGURE 2.8   Variation of loss tangent with frequency, showing maxima (Maxwell-Wagner effect). The increase in loss tangent is smaller when SiO_2 is replaced with air.

2.3   FABRICATION OF AIR-CLAD COPPER INTERCONNECTS

The fabrication of air-clad interconnects has been demonstrated by several research groups. Each new design offers a different fabrication protocol with specific limitations. Some groups have used a thermally degradable polymer (TDP) that decomposes to form an air cavity in specific regions [5,6,10], while others use a nonconformal chemical vapor deposition (CVD) process of the dielectric material, such as SiO_2 [3,5,7] or SiCN [18]. The nonconformal deposition creates trapped voids in specific regions. It is difficult to incorporate a TDP in a multilevel metallization module (MLM); hence sacrificial polymers have been generally used in chip-to-board transmission lines [5,19]. Several research groups have also used a wet/dry etching technique to remove the sacrificial dielectric material and create air gaps in stacked structures [24,26,27]. Typically, for several layers of air-clad interconnects, the process is either a buildup of trapped air pockets through nonconformal deposition or wet/dry etching of a sacrificial material either after every metal level or after the buildup is complete. The different processing techniques used for air-gap fabrication of interconnects are now reviewed.

Sukharev et al. [14] proposed a nonconformal deposition technique to entrap voids using a two-step deposition process. After the pillar metal interconnects are fabricated on a silicon wafer, SiO_2 is deposited by plasma enhanced chemical vapor deposition (PECVD) using tetraethoxysilane (TEOS) as the reactant, which gives a conformal layer and hence a uniform topography. The thickness of this conformal PECVD–TEOS layer determines the sidewall thickness of the air gap. The second step involves PECVD deposition using silane and oxygen reactants resulting in a nonconformal layer that creates an air gap as shown in Figure 2.9.

Trapped air gaps can affect the electromigration kinetics primarily through changes in the effective rigidity of the confinement materials [14]. Air-gap structures offer lower mechanical stability, which in turn results in a lower residual stress due to the depletion or accumulation of the electromigrating atoms. B. P. Shieh et al. [15] used an additional HDP–CVD layer to limit the air-gap extension beyond the confined region, as shown in Figure 2.10. The HDP–CVD process resputters oxide material, thus confining the air gap by restricting its extension above the metal lines.

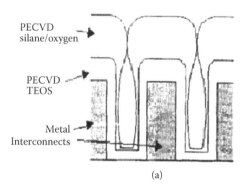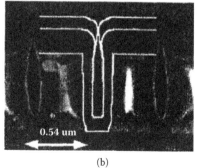

(a)   (b)

FIGURE 2.9   (a) Schematic of trapped void and (b) SEM image of the air-cavity with SPEEDIE simulation profiles. (From V. Sukharev et al., *Microelectronic Reliability*, vol. 41, pp. 1631–1635, 2001. With permission.)

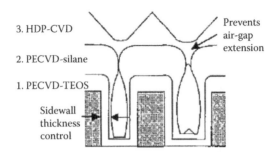

FIGURE 2.10 Schematic of air-gap interconnect process. (© 2002 IEEE. Reprinted with permission.)

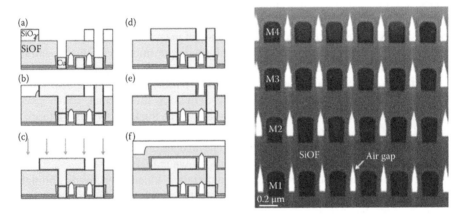

FIGURE 2.11 (Left) Process layout in the fabrication of air-gap interconnects: (a) via/trench DD process, (b) Cu deposition (Cu CMP), (c) etchback, (d) wet cleaning (H_2 annealing), (e) SiON deposition, and (f) dielectric deposition (ILD CMP). (Right) TEM image shows four levels of air-gap interconnects. (© 2009 IEEE. Reprinted with permission.)

Noguchi et al. [8] fabricated four levels of air-gap interconnects using a similar process sequence. These structures were fabricated using a dual damascene approach followed by an etchback process. The etchback process incorporated regions that form trapped air gaps during deposition of the interlayer dielectric material, as shown in Figure 2.11. The air gaps were selectively formed between adjacent wires with a minimum pitch of 180 nm lines and spaces. The air gaps formed through this process tended to have a distinct shape like that of the via base. It was shown that the air-gap interconnect has the same via electromigration lifetime as a damascene produced interconnect. The air-gap-based interconnect also tends to have a better time dependent dielectric breakdown (TDDB) reliability compared to the damascene process.

Nakamura et al. [18] proposed a novel process to fabricate air gaps in multilevel architecture in a single step. They use a gas composition to dry-etch certain sacrificial regions. First, a hybrid dual damascene structure using polyarylene ether (PAr) with SiOC was fabricated where PAr was used as a trench layer material and the SiOC film

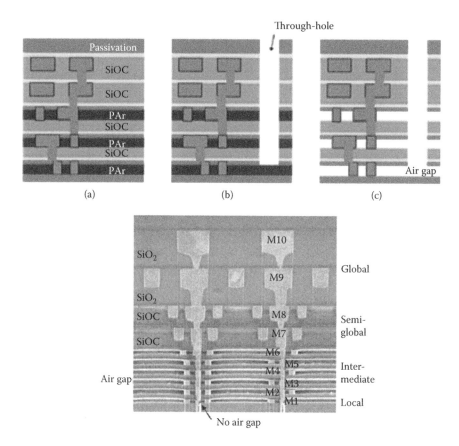

FIGURE 2.12 Schematic image of the air-gap process. The bottom image is an SEM image of the 10-level interconnect structure showing air gaps. (© 2008 IEEE. Reprinted with permission.)

was used as a via layer material. Once several intermediate and semiglobal layers were fabricated, a through-hole was formed in places where the PAr material was removed (Figure 2.12b). The gas was composed of O_2-N_2-H_2 and introduced at the in-let through holes. The PAr regions were etched away leaving air gaps (Figure 2.12c). In order to control the moisture uptake and seal the gas outlet once the air gap was fabricated, the holes were refilled using a spin-on dielectric (SOD). The viscosity of the SOD can be changed according to the degree of refill required. The excess SOD was removed using a dry etchback process and a CVD–SiCN film was deposited as a hermetic seal [18].

Yoo et al. [23] recently demonstrated air gaps through nonconformal, PECVD deposition to form the interlayer dielectric between metal lines. The dielectric was first removed and then the nonconformal PECVD material formed the air gaps. A two-step deposition was carried out including a conformal dielectric deposition followed by a nonconformal, low-k interlayer dielectric material. The final air gap shape depended on the characteristics of the interlayer dielectric removal process, and the conformal nature of the PECVD material, as shown in Figure 2.13.

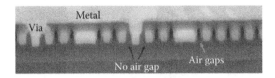

FIGURE 2.13   SEM image of air gaps. (© 2010 IEEE. Reprinted with permission.)

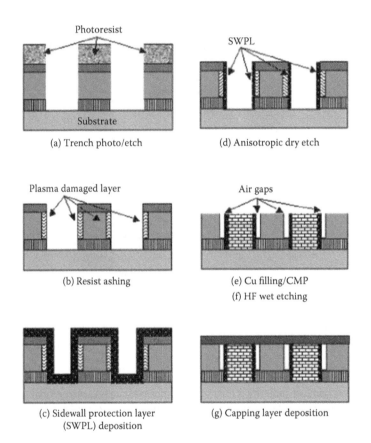

FIGURE 2.14   Schematic of self-aligned air-gap structure formed by wet etching. (© 2008 IEEE. Reprinted with permission.)

Photolithography was used as a masking step before etching, which allowed precise control of the air gap placement between closely spaced metal lines where the electrical coupling was the highest. Conventional 193 nm dry lithography was used to pattern the air-gap features for critical layers in 32 nm and 22 nm technology node vehicles. H.-W. Chen et al. [24] proposed a novel wet-etching technique to create a self-aligned air-gap interconnect. A layer of plasma-damaged SiOC was first formed on the sidewalls of a standard Cu/SiOC (low k) damascene structure, as shown in Figure 2.14. Then, a thin layer of dielectric material was deposited and etched back to form a sidewall protection layer (SWPL). Air gaps were created after copper filling and CMP by selectively removing the damaged SiOC layer through wet etching with HF. Finally, the structure was capped with a dielectric layer. Without the SWPL, the dielectric layer below metal lines could be inadvertently undercut during prolonged wet etching causing damage to the metal lines. The wet etching step puts a limit on the air-gap size due to process constraints.

The SWPL limits the amount of capping layer deposited into the air gaps, so that the shape and size of air gaps are better retained. The capping dielectric is confined at the top of the air gap and serves as an extra etch stop. The authors showed that these air gaps increase the misalignment tolerance of the via. A doubling of the electromigration resistance was observed due to an increase in back stress. Similar sidewall air-gap integration has also been shown by Gueneau de Mussy et al. [25].

Gras et al. [26] incorporated air gaps into a multilayered metal stack using wet etching of SiO_2. The metal-1/metal-2 (M1-M2) copper interconnects were first fabricated, as shown in Figure 2.15, using a trench-first hard mask integration scheme in SiO_2. Line capping was done on the M1 layer using a CoWP self-aligned barrier to allow HF diffusion at the lower level during the etching process. A thin layer of SiCN was deposited on the top of M2 to localize the air-cavities in the global structure. Subsequently, a noncritical mask was developed to open the apertures in the SiCN in order to generate localized HF diffusion pathways. The SiCN is not attacked by HF during the process. The HF progresses isotropically from these apertures into the underlying SiO_2 layers resulting in the removal of SiO_2. After the fabrication of the air gaps in specific locations, the upper level intermetal dielectric (IMD) is deposited, closing the SiCN apertures. It was observed that a substantial improvement in RC delay resulted from the creation of air gaps. However, there was a slight degradation in electromigration resistance due to line resistance increase from the formation of voids.

Anand et al. [27] proposed the gas-dielectric interconnect process (NURA), which uses the decomposition of carbon to form air pockets in interconnect architecture. In this process, the trenches in the interconnect metal were first formed in carbon instead of an insulator, as shown in Figure 2.16. A thin layer of SiO_2 was then deposited. A standard dual damascene process was incorporated into this structure to form a dual NURA process. Subsequently, furnace ashing was carried out where the structure was heated to 400°C to 450°C in oxygen. The oxygen diffuses through the SiO_2 and reacts with the carbon leaving the wire-to-wire spaces filled with a gas. These gas-filled structures were observed to cause a 50% reduction in delay. However, there are certain difficulties in this process including higher temperatures to oxidize carbon, the potential oxidation of the metal, and need to permeate oxygen through to the inner layers.

Substantial work has been carried out with thermally degradable polymer to create air cavities for low-loss transmission lines. Incorporating a sacrificial polymer, such as a polycarbonate or polynorbornene, in a multilevel metal stack is challenging since the introduction of a new material into the MLM interconnect stack faces integration challenges with respect to patterning control and dimensional shrinkage. In addition, this approach does not allow local integration of air gaps, due to mechanical issues during packaging steps when large air gaps are introduced [26].

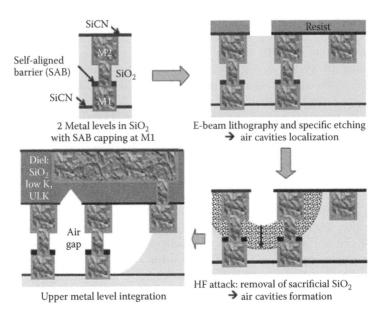

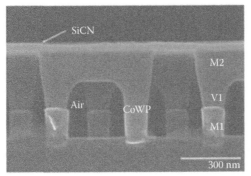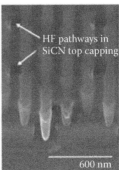

FIGURE 2.15 (Top) Schematic of air-gap fabrication process through wet etching of oxide. (Bottom) SEM images of fabricated air gaps in MLM stack. (© 2008 IEEE. Reprinted with permission.)

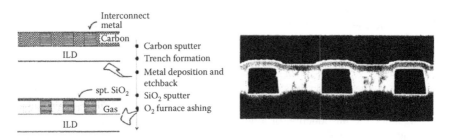

FIGURE 2.16 (Left) Schematic of the single-NURA process and (right) SEM image of the fabricated air gaps. (© 1996 IEEE. Reprinted with permission.)

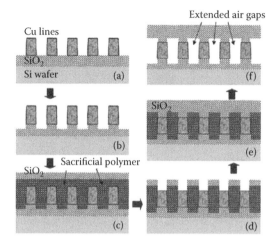
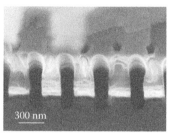

FIGURE 2.17 (Left) Schematic of the process flow for extended air-gap structure. (Right) SEM image of the air gaps. (From S. Park et al., "Materials, processes, and characterization of extended air-gaps for the intra-level interconnection of integrated circuits," Ph.D. dissertation, Georgia Institute of Technology, 2008. With permission.)

Hence, most of the ongoing research is focused on fabricating novel structures to enable low-loss, chip-to-chip connectivity at the board level.

Park et al. [6] demonstrated an extended air-gap structure using tetracyclododecene (TD)-based sacrificial material. Polymethyl-methacrylate (PMMA) was first spin coated and patterned using e-beam lithography, as shown in Figure 2.17. A Cu layer was first sputtered followed by deposition of PMMA. PMMA was removed leaving copper lines on the SiO_2 dielectric. The oxide was etched back to a thickness of 100 nm by reactive ion etching. The sacrificial polymer was spin coated and a thin SiO_2 layer was deposited. The polymer was directionally etched from top. The copper lines were patterned using electron beam lithography. Finally, the upper SiO_2 layer was deposited and the sacrificial polymer was thermally decomposed to form the air gaps, as shown in Figure 2.17. The effective dielectric constant of this structure was measured to be 2. Similar structures were also shown by Kohl et al. [10].

A different transmission line structure using polycarbonate sacrificial materials has been recently demonstrated [29]. The signal and ground lines were fabricated on an FR4 board using polypropylene carbonate as the sacrificial material between signal and ground lines, as shown in Figure 2.18. First, a copper seed layer was deposited by DC sputtering on an FR4 board with a thin (~150A°) layer of titanium to improve adhesion of the copper to the substrate. A positive tone photoresist was used to pattern the ground lines. These ground lines were subsequently electroplated using a photoresist mold. The electroplating conditions contribute to the surface roughness and thickness of the copper lines, which affects the transmission line characteristics.

The sacrificial polypropylene carbonate-based polymer was then spun onto the structure and patterned. Usually, a photoacid generator was added to the polypropylene carbonate to make it photopatternable. The signal lines were then sputter

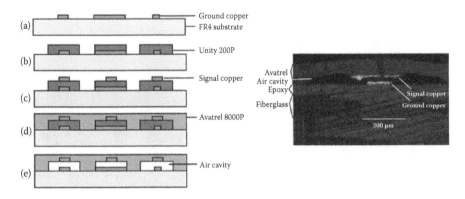

FIGURE 2.18 (Left) Flow diagram for air-cavity clad interconnects using sacrificial polymer. (Right) An optical image of the air-cavity structure. (© 2012 IEEE. Reprinted with permission.)

coated and electroplated to full thickness on top of the sacrificial material, which was subsequently patterned. The photopatternable overcoat material was spin coated onto the top of the signal lines. Usually, an adhesion promoter was used to improve the adhesion strength between the overcoat and the copper signal lines. After the overcoat was patterned, the sacrificial material was decomposed through the overcoat to form the air cavity. For conventional FR4 boards, decomposition temperatures must be restricted to 200°C to prevent board damage. Silicon substrates allow for higher processing temperatures compared to FR4 boards; however, the decomposition temperature should not thermally damage the overcoat material. Once the ground and signal line structures were fabricated, the chip-to-board connectivity could require extra processing steps. The chip attachment could be completed using flip-chip solder connections. Chen et al. [19] further demonstrated an active low-power link with a transmitter and receiver using this air-gap structure. A 26% reduction in dielectric loss was realized. The active link was operated at 6.25 Gb/s and consumed 3.7 mW using a 1.2 V power supply.

A slightly different structure was demonstrated by Spencer et al. [5]. In that disclosure, the overcoat material was placed between the ground and signal lines such that the ground line rested on the overcoat, as shown in Figure 2.19. This structure does not have a complete air cavity between signal and ground, and thus does not take full advantage of the air gap, however, it may be more robust mechanically.

Novel hybrid designs of TDP air cavities have been proposed and are currently being investigated [21]. One such structure has been shown in Figure 2.4 and its process flow has been shown in Figure 2.20. These structures incorporate footprints for Twinax cables to achieve complete off-chip link. As shown in Figure 2.20, copper ground lines are first electroplated to desired thickness and subsequently patterned. A thick polyimide layer is spun and patterned on top of copper and the thermally degradable polymer, PPC, is patterned between the polyimide walls. A thin oxide/polyimide support is built on top of the polyimide and the PPC is decomposed to reveal the air cavity. A differential signal pair (copper) is patterned and electroplated on top of the air gap supported by the thin oxide/polyimide interface.

FIGURE 2.19   SEM image of air-gap interconnect with ground line resting on the overcoat. (© 2007 IEEE. Reprinted with permission.)

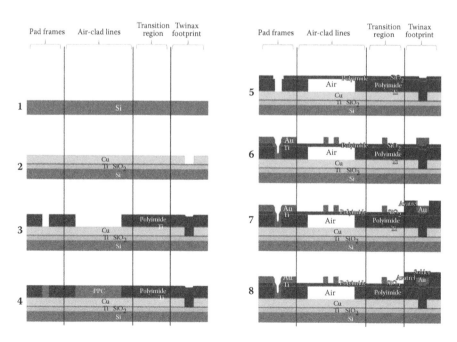

FIGURE 2.20   Proposed process flow for the hybrid air-cavity structure that incorporates Twinax footprints for off-chip connections.

The Twinax footprint is then fabricated with Avatrel 8000P as the mold for the solder connections that are electroplated on sputtered gold.

There are several additional processing steps required in such structures to incorporate the Twinax connection. For vertical integration, another novel technique has been recently proposed that uses TDPs to form air clad through TSVs (Figure 2.5). The extension of air-clad transmission lines to impedance-matched, air-clad TSVs can provide significant electrical improvement in the overall structure. The proposed fabrication flow has been listed in Figure 2.21. An electroless process is first used to coat the TSV sidewalls with copper. Then the TSVs are filled with PPC and the PPC is subsequently decomposed. Due to the unique interaction between PPC solution and copper on the sidewall of TSV, the thermal decomposition temperature of PPC in contact with copper is increased [30]. Hence, a thin layer of PPC coats the copper sidewalls after the PPC decomposition. The copper via is electroplated through the silicon and the PPC coat is decomposed at higher temperatures to reveal the air-clad connection. Electroless copper has poor adhesion to silicon and this is one of the main limitations of this process. Removal of seed layer of electroless copper will require mechanical polishing on both sides of the wafer.

In each of the techniques that have been discussed, there are specific advantages and drawbacks that affect the integration of these advances into a CMOS process flow. The main drawback of the nonconformal deposition methods is that they require an additional air-gap lithography step at each metal level. Since air gaps may be introduced inside the narrowest spaces where the highest electrical performance is required, the cost of additional lithography steps has to be considered [27]. To overcome this limitation, blanket dielectric etchback techniques (direct etching of trenches), which have minimal damage to the copper lines, are under investigation.

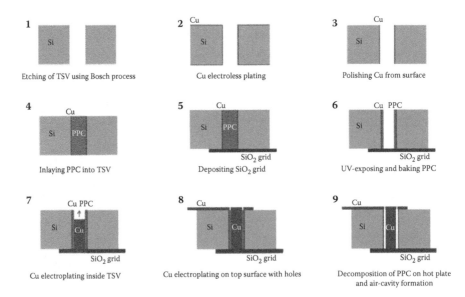

FIGURE 2.21 Process flow for air-clad TSVs.

The use of polymers as sacrificial materials poses several constraints as well. It is difficult to implement a complete damascene integration scheme at 200°C. The mechanical stability of the IMD after air-gap formation poses a question. If the sacrificial material were removed after each metal level, the issue of via misalignment needs to be addressed [27]. Due to these constraints, TDPs have primarily been used for board-level interconnects.

2.4 CONCLUSION

Air-gap structures have been shown to reduce the line capacitance by reduction in the effective dielectric constant. However, the applications in chip-to-chip connectivity are a relatively new concept. Chip-to-chip interconnects represent a major bottleneck in the performance of high-speed electronic systems. Air-clad, chip-to-chip interconnects can offer significant performance improvements over conventional substrates. Results have shown significant reduction in dielectric loss by using these air-clad interconnects. Also, the use of low-k materials with air dielectrics require a redesign of the metal and dielectric spacing ensuring that the interconnect lines need to be wider for the same impedance value. This results in further lowering of the conductor losses. Together, these two phenomena lead to reduced energy/bit and higher bandwidth density. Thus, significant improvements can be realized with planar links using air-clad interconnects. However, vias, discontinuities, and transitions continue to remain performance bottlenecks for chip-to-chip links in 3D interconnect and requires continued research emphasis. From a processing standpoint, both nonconformal deposition of dielectric through PECVD process as well as etching/decomposition of sacrificial layers show promise. However, specific challenges need to be addressed.

ACKNOWLEDGMENTS

The authors acknowledge the support of the Interconnect Focus Center, one of six research centers funded under the Focus Center Research Program, a Semiconductor Research Corporation program. The authors also gratefully acknowledge the valuable technical discussions and guidance received from the members of the Naeemi Research Group at Georgia Institute of Technology and the Bashirullah Research Group at University of Florida.

REFERENCES

1. W. J. Dally and J. W. Poulton, *Digital Systems Engineering*, Cambridge University Press, Cambridge, UK, 1998.
2. R. Sharma and T. Chakravarty, *Compact Models and Measurement Techniques for High-Speed Interconnects*, Springer, New York, 2012.
3. B. Shieh et al., "Air-gap formation during IMD deposition to lower interconnect capacitance," *IEEE Electron Device Letters*, vol. 19, no. 1, pp. 16–18, 1998.
4. B. P. Shieh et al., "Integration and reliability issues for low capacitance air-gap interconnect structures," *Proceedings of IITC 1998*, pp. 125–127, 1998.

5. T. J. Spencer et al., "Air-gap transmission lines on organic substrates for low-loss interconnects," *IEEE Transactions on Microwave Theory and Techniques*, vol. 55, no. 9, pp. 1919–1925, 2007.

6. S. Park et al., "Materials, processes, and characterization of extended air gaps for the intra-level interconnection of integrated circuits," Ph.D. dissertation, Georgia Institute of Technology, 2008.

7. J. Noguchi et al., "Simple self-aligned air-gap interconnect process with Cu/FSG structure," *Proceedings of IEEE IITC 2003*, pp. 68–70, 2003.

8. J. Noguchi et al., "Multilevel interconnect with air-gap structure for next generation interconnections," *IEEE Transactions on Electron Devices*, vol. 56, no. 11, pp. 2675–2682, 2009.

9. L. G. Gosset et al., "Integration of SiOC air gaps in copper interconnects," *Microelectronics Engineering*, vol. 70, no. 2–4, pp. 274–279, 2003.

10. P. A. Kohl et al., "Air gaps for electrical interconnections," *Electrochemical and Solid-State Letters*, vol. 1, no. 1, pp. 49–51, 1998.

11. P. A. Kohl et al., "Air-gaps in 0.3 μm electrical interconnections," *IEEE Electron Device Letters*, vol. 21, no. 12, 2000.

12. S. Park et al., "Air gaps for high-performance on-chip interconnect, Part I: Improvement in thermally decomposable template," *Journal of Electronic Materials*, vol. 37, no. 10, pp. 1524–1532, 2008.

13. T. J. Spencer et al., "Air cavity low-loss signal lines on BT substrates for high-frequency chip-to-chip communication," *Proceedings of IEEE ECTC 2009*, pp. 1221–1226, 2009.

14. V. Sukharev et al., "Reliability studies on multiple interconnection with intermetal dielectric air gaps," *Microelectronic Reliability*, vol. 41, pp. 1631–1635, 2001.

15. B. P. Shieh et al., "Electromigration reliability of low capacitance air-gap interconnect structures," *Proceedings of IEEE IITC 2002*, pp. 203–205, 2002.

16. M. Pantouvaki et al., "Dielectric reliability of 70 nm pitch air-gap interconnect structures," *Microelectronic Reliability*, vol. 88, pp. 1618–1622, 2011.

17. X. Zhang et al., "Impact of process induced stresses and chip-packaging interaction on reliability of air-gap interconnects," *Proceedings of IEEE IITC 2008*, pp. 135–137, 2008.

18. N. Nakamura et al., "Cost-effective air-gap interconnects by all-in-one post-removing process," *Proceedings of IEEE IITC 2008*, pp. 193–195, 2008.

19. J. Chen et al., "Air-cavity low-loss transmission lines for high-speed link applications," *Proceedings of IEEE ECTC 2011*, pp. 2146–2151, 2011.

20. V. Kumar et al., "Modeling, optimization and benchmarking of chip-to-chip electrical interconnects with low-loss air-clad dielectrics," *Proceedings of IEEE ECTC 2011*, pp. 2084–2090, 2011.

21. R. Sharma et al., "Design and fabrication of low-loss horizontal and vertical interconnect links using air-clad transmission lines and through silicon vias," *Proceedings of IEEE ECTC 2012*, pp. 2005–2012, 2012.

22. H. Hasegawa et al., "Properties of microstrip line on Si-SiO₂ system," *IEEE Transactions on Microwave Theory and Techniques*, vol. 19, no. 11, pp. 869–881, 1971.

23. H. J. Yoo et al., "Demonstration of a reliable high-performance and yielding air gap interconnect process," *Proceedings of IEEE IITC 2010*, pp. 1–3, 2010.

24. H.-W. Chen et al., "A self-aligned air-gap interconnect process," *Proceedings of IEEE IITC 2008*, pp. 34–36, 2008.

25. J. P. Gueneau de Mussy et al., "Selective sidewall air-gap integration for deep submicrometer interconnects," *Electrochemical and Solid-State Letters*, vol. 7, no. 11, pp. G286–G289, 2004.

26. R. Gras et al., "300-mm multilevel air-gap integration for edge interconnect technologies and specific high-performance applications," *Proceedings of IEEE IITC 2008*, pp. 196–198, 2008.

27. M. B. Anand et al., "NURA: A feasible gas-dielectric interconnect process," *1996 Symposium on VLSI Technology Digest of Technical Papers*, pp. 82–83, 1996.
28. L. G. Gossett et al., "General review of issues and perspectives for advanced copper interconnections using air gap as ultra-low K material," *Proceedings of IEEE IITC 2003*, pp. 65–67, 2003.
29. Todd J. Spencer et al., "Air-cavity transmission lines for off-chip interconnects characterized to 40GHz," *IEEE Transactions on Components, Packaging and Manufacturing Technology*, vol. 2, no. 3, pp. 367–374, 2012.
30. T. Spencer et al., "Stabilization of the thermal decomposition of poly(propylenecarbonate) through copper ion incorporation and use in self-patterning," *Journal of Electronic Materials*, 40, p. 1350, 2011.

3 Silicon Photonic Bragg Gratings

Xu Wang, Wei Shi, and Lukas Chrostwoski

CONTENTS

3.1 INTRODUCTION

Silicon photonics, with great potential for bringing together two technological areas that have transformed the last century—electronics and photonics—is gaining significant momentum because it allows photonic devices to be made cheaply using complementary metal-oxide semiconductor (CMOS) fabrication techniques and integrated with electronic chips. The Bragg grating, a fundamental component for achieving wavelength selective functions, has been used in numerous optical devices such as semiconductor lasers and fibers. Over the past few years, the integration of Bragg gratings in silicon waveguides has been attracting increasing research interest. In this chapter, we discuss the recent development of silicon photonic Bragg gratings, starting from the simple uniform gratings to applications in more complex grating devices. We also provide insight into some practical issues and challenges involved with the design and fabrication.

3.2 UNIFORM BRAGG GRATINGS IN SILICON WAVEGUIDES

3.2.1 THEORY

In the simplest configuration, a Bragg grating is a structure with periodic modulation of the effective refractive index in the propagation direction of the optical mode, as shown in Figure 3.1. This modulation is commonly achieved by varying the refractive index (e.g., alternating material) or the physical dimensions of the waveguide. At each boundary, a reflection of the traveling light occurs, and the relative phase of the reflected signal is determined by the grating period and the wavelength of the light. The repeated modulation of the effective index results in multiple and distributed reflections. The reflected signals only interfere constructively in a narrow band around one particular wavelength, namely the Bragg wavelength. Within this range, light is strongly reflected. At other wavelengths, the multiple reflections interfere destructively and cancel each other out, and as a result, light is transmitted through the grating. Figure 3.2 shows the typical spectral response of a uniform Bragg grating. The Bragg wavelength is given as

$$\lambda_B = 2 \Lambda n_{\text{eff}} \tag{3.1}$$

where Λ is the grating period and n_{eff} is the average effective index. Based on coupled-mode theory [1], the reflection coefficient for a uniform grating with length L is described by

$$r = \frac{-i\kappa \sinh(\gamma L)}{\gamma \cosh(\gamma L) + i \Delta\beta \sinh(\gamma L)} \tag{3.2}$$

with

$$\gamma^2 = \kappa^2 - \Delta\beta^2 \tag{3.3}$$

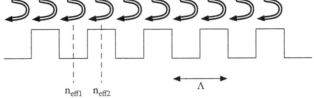

n_{eff1} n_{eff2}

Λ

FIGURE 3.1   Longitudinal effective index profile of a uniform grating.

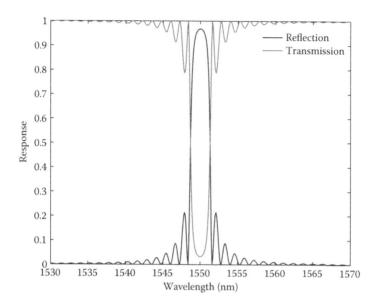

FIGURE 3.2 Typical spectral response of a uniform Bragg grating.

Here, $\Delta\beta$ is the propagation constant offset from the Bragg wavelength,

$$\Delta\beta = \beta - \beta_0 \ll \beta_0 \tag{3.4}$$

and κ is often defined as the coupling coefficient of the grating and can be interpreted as the amount of reflection per unit length. For a stepwise effective index variation as shown in Figure 3.1 ($\Delta n = n_{\text{eff2}} - n_{\text{eff1}}$), the reflection at each interface can be written as $\Delta n / 2 n_{\text{eff}}$ according to the Fresnel equations. Each grating period contributes two reflections, therefore the coupling coefficient is

$$\kappa = 2 \frac{\Delta n}{2 n_{\text{eff}}} \frac{1}{\Lambda} = \frac{2\Delta n}{\lambda_B} \tag{3.5}$$

For a sinusoidal effective index variation $n(z) = n_{\text{eff}} + \Delta n / 2 \cdot \cos(2\beta_0 z)$, the coupling coefficient is given as [1]

$$\kappa = \frac{\pi \Delta n}{2 \lambda_B} \tag{3.6}$$

Similarly, for other effective index variations, we can take the Fourier expansions, $n(z) = n_{\text{eff}} + \sum_i \Delta n_i / 2 \cdot \cos(i \cdot 2\beta_0 z)$, and the coupling coefficient can be derived from the first-order Fourier component, $\kappa = \pi \Delta n_1 / (2 \lambda_B)$.

For the case where $\Delta\beta = 0$, Equation (3.2) is written as $r = -i\tanh(\kappa L)$, therefore, the peak power reflectivity at the Bragg wavelength is

$$R_{\text{peak}} = \tanh^2(\kappa L) \tag{3.7}$$

The bandwidth is also an important figure of merit for Bragg gratings. The bandwidth between the first nulls around the resonance can be determined by [1]

$$\Delta\lambda = \frac{\lambda_B^2}{\pi n_g}\sqrt{\kappa^2 + (\pi / L)^2} \qquad (3.8)$$

where n_g is the group index. It should be noted that this is larger than the 3 dB bandwidth.

3.2.2 INTEGRATION IN SILICON WAVEGUIDES

The integration of Bragg gratings in silicon-on-insulator (SOI) waveguides was first demonstrated by Murphy et al. [2] in 2001. Typically, the gratings are achieved by physically corrugating the silicon waveguides. This is in contrast to the manufacture of fiber Bragg gratings, in which the fiber is photosensitive and exposed to intense ultraviolet (UV) light to induce refractive index modulation in the core. Instead of using physical corrugations, there are a few other approaches to form gratings in silicon, such as ion-implanted Bragg gratings [3], however, they are less common and thus beyond the scope of this text.

In this section, we will discuss two waveguide structures that are most commonly used for integrated Bragg gratings: strip waveguides and rib waveguides.

3.2.2.1 Strip Waveguide Gratings

Strip waveguides (also referred to as photonic wires and channel waveguides) usually have very small cross-sections. Figure 3.3 shows the cross-section of a strip waveguide that is widely used in state-of-the-art silicon photonic circuits. The thickness of the top silicon layer and the buried oxide layer is 220 nm and 2 μm, respectively, and the waveguide width is 500 nm. As shown in Figure 3.4, light is strongly confined in the core, due to the high refractive index contrast between the core (silicon) and the cladding (oxide or air). This strong optical confinement allows for very tight bends (bending radius can be a few microns and the bending loss is negligible).

The grating corrugations are normally on the waveguide sidewalls, therefore, the grating and the waveguide can be defined in a single lithography step. Due to

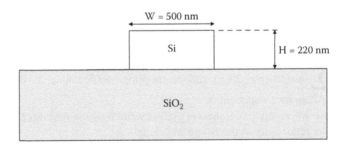

FIGURE 3.3 Schematic diagram of the cross-section of a strip waveguide.

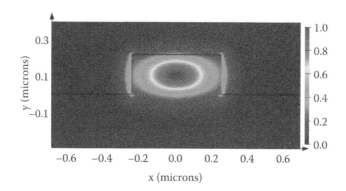

FIGURE 3.4 Simulated fundamental transverse-electric (TE) mode profile in the strip waveguide.

the small waveguide geometry and optical mode size, a small perturbation on the sidewalls can cause a considerable grating coupling coefficient, thus resulting in a large bandwidth. The experimentally demonstrated bandwidth is generally on the order of tens of nanometers [4,5]. The lowest bandwidth reported so far is about 0.8 nm [6], however, the authors used a very small corrugation width of 10 nm in the design, and the actually fabricated corrugations were even smaller due to the lithography smoothing effect. Therefore, it is quite challenging to fabricate such small corrugations directly on the sidewalls. Tan et al. demonstrated another concept by moving the sidewall corrugations outside of the waveguide and placing a periodic array of cylinders near the waveguide to achieve similarly small effective index perturbations [7]. The reported bandwidth using this approach is on the order of a few nanometers; however, this approach is still sensitive to fabrication errors because the cylinders are still small (200 nm diameter) and are isolated structures.

Figure 3.5a shows the tilted cross-section SEM image of a fabricated straight strip waveguide [6]. It should be noted that the cross-section profile is not perfectly rectangular and has slightly sloped sidewalls. Furthermore, the waveguide width and thickness also slightly deviate from the design values. Such geometric imperfections will affect the effective index of the waveguide, and, as a result, usually shift the Bragg wavelength from its design value. Figure 3.5b shows the top view SEM image of a fabricated Bragg gratings. It is important to keep in mind that if using optical lithography in the fabrication, the gratings actually fabricated can be severely smoothed. This is the case in Figure 3.5b, where square corrugations were used in the mask design, but the fabricated corrugations are severely rounded and resemble sinusoidal shapes. Therefore, to obtain the desired bandwidth, such smoothing effects should be taken into account and compensated. This can be achieved by simply using larger corrugation widths in the mask design than in the simulation or by adding mask correction features [9]. However, this cannot be easily done without lithography simulations [10], which will be discussed later.

Figure 3.6 shows the measured transmission spectra for the strip waveguide gratings, where the corrugation widths are design values. As the corrugation width is increased, the coupling coefficient increases, leading to broader bandwidth.

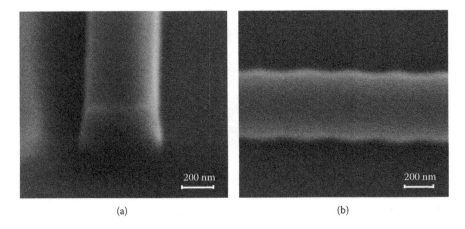

(a)   (b)

FIGURE 3.5 SEM images of strip waveguide gratings: (a) cross-section of a straight strip waveguide and (b) top view of the gratings. (After Xang et al., *Opt. Express*, vol. 20, no. 14, pp. 15547–15558, 2012.)

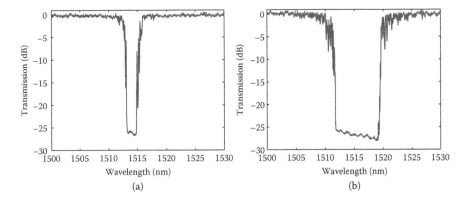

FIGURE 3.6 Measured transmission spectra of strip waveguide gratings: (a) 10 nm corrugations, (b) 40 nm corrugations.

It should be noted that the bandwidth using 10 nm corrugations, as shown in Figure 3.6a, is about 1.6 nm, is almost twice the reported value from Wang et al. [6]. The corrugations were fabricated in the same foundry but in different runs. This indicates that the such small corrugations are highly sensitive to fabrication variations even using a relatively mature process [11].

3.2.2.2 Rib Waveguide Gratings

As discussed earlier, strip waveguide gratings have relatively large bandwidths and are sensitive to fabrication variations. However, numerous applications require narrow bandwidths, such as wavelength-division multiplexing (WDM) channel filters. An alternative is to use rib waveguides, which typically have larger cross-sections and allows higher fabrication tolerance.

There are several configurations to form gratings on rib waveguides. The grating corrugations can be the top surface [2,12,13] or the sidewalls, where the sidewalls can be corrugated either on the rib [14] or on the slab [15]. The top-surface-corrugated configuration usually has a fixed etch depth, therefore, the grating coupling coefficient is constant. The sidewall-corrugated configuration has the advantage that the corrugation width can be easily controlled, which is essential to make complex grating profiles, such as apodized gratings that can suppress reflection side lobes [15].

The rib waveguide geometry is often designed to be single mode [16]. However, a nominally single-mode rib waveguide can have higher-order leaky modes, which cause unwanted dips in the transmission spectrum on the shorter wavelength side of the fundamental Bragg wavelength. To separate these leaky modes away from the fundamental mode, it is necessary to shrink the waveguide dimensions [2]. This is the general trend in silicon photonics as well, because small waveguide dimensions are desired for high integration density and improved cost efficiency. However, most integrated Bragg gratings were demonstrated in rib waveguides with relatively large cross-sections until our recent work [8].

Figure 3.7 illustrates the waveguide structure [8]. The only difference from the strip waveguide shown in Figure 3.3 is the extra slab region. The silicon thickness (H) is still 220 nm, and the shallow etch depth (D) is 70 nm. The rib width (W_1) is 500 nm, and the slab width (W_2) is only 1 μm. As shown in Figure 3.8, most light is confined under the rib and the optical field's overlap with the sidewalls is very low around both the rib and slab sidewalls. This overlap reduction makes it possible to introduce weaker effective index perturbations compared to the strip waveguide gratings, thus allowing for smaller coupling coefficients and narrower bandwidths. In addition, the propagation loss is also reduced. The gratings are realized by introducing periodic sidewall corrugations either on the rib or on the slab. Figure 3.9 shows the SEM images of the fabricated devices.

The device layout schematic is also shown in Figure 3.10. The input and output ports are waveguide-to-fiber grating couplers [17] that are designed for transverse-electric (TE) polarization. A Y-branch splitter was placed between the input grating coupler and the rib waveguide gratings to collect the reflected light. The strip waveguide illustrated in Figure 3.3 is used for the routing waveguides as well as for the Y-branch splitter in order to minimize their footprints and bending losses.

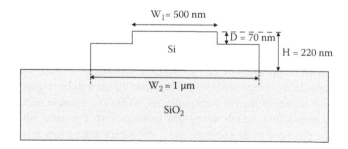

FIGURE 3.7   Schematic diagram of the cross-section of the rib waveguide.

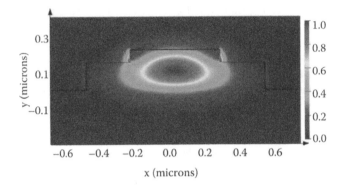

FIGURE 3.8 Simulated fundamental transverse-electric (TE) mode profile in the rib waveguide.

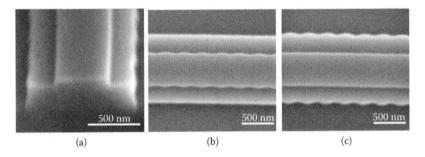

FIGURE 3.9 SEM images of rib waveguide gratings: (a) cross-section of a rib waveguide with gratings on the slab; (b) top view of gratings on the rib: corrugation width is 60 nm (design value); and (c) top view of gratings on the slab: corrugation width is 80 nm (design value). (After Xang et al., *Opt. Express*, vol. 20, no. 14, pp. 15547–15558, 2012.)

Double-layer linear tapers are also designed for the transition between the strip and rib waveguides, and are long enough to ensure that the transition loss is negligible.

Figure 3.11 shows the measured transmission and reflection spectra for a fabricated rib waveguide gratings designed with 60 nm corrugations. It can be seen that only one dip exists in the transmission spectrum in a wide wavelength range, indicating that the higher-order leaky modes are far away from the fundamental mode (thus can be ignored). More important, the bandwidth is only about 0.8 nm, which is actually smaller than that of the strip waveguide grating using 10 nm corrugations shown in Figure 3.6a.

Figure 3.12 plots the experimental bandwidth as a function of the designed corrugation width for devices on one die. To obtain a subnanometer bandwidth, if using the strip waveguide, a corrugation width of less than 10 nm will be required, whereas, if using the slab region of the rib waveguide, an 80 nm corrugation width should be suitable. This means that the rib waveguide gratings have a relaxed fabrication tolerance. For the rib waveguide gratings, the 3 dB bandwidth ranges from 0.4 nm to 0.8 nm, which is suitable for many narrow-band applications such as WDM channel filters, although apodization may be required to suppress the side lobes.

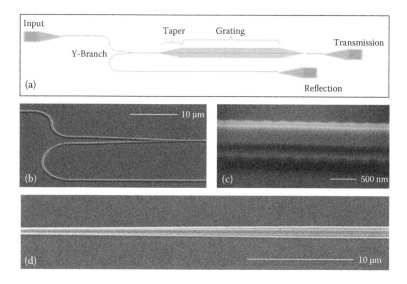

FIGURE 3.10   (a) Schematic diagram of the chip layout, and SEM images of (b) Y-branch, (c) gratings, and (d) strip-to-rib taper. (After Wang et al., *Opt. Express*, vol. 20, no. 14, pp. 15547–15558, 2012.)

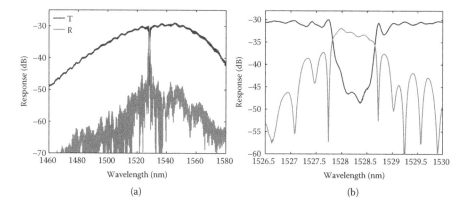

FIGURE 3.11   (a) Measured spectral responses of a fabricated rib waveguide grating designed with 60 nm corrugations on the rib. The spectra are plotted without normalization. (b) Enlarged plot around the Bragg wavelength.

In summary, the rib waveguide is optimized to reduce the optical fields around the sidewalls, while keeping the cross-section small and maintaining single-mode operation. Both configurations show much narrower bandwidths and higher fabrication tolerances than the strip waveguide gratings. Besides, no higher-order leaky modes are observed in a very wide spectral range. This is a significant advantage over ring resonators with relatively small free spectral ranges and gratings on bulky rib waveguides with higher-order leaky modes.

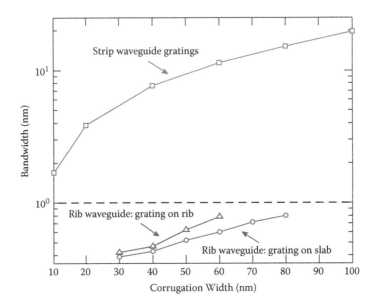

FIGURE 3.12   Measured 3 dB bandwidth versus the designed corrugation width for different grating structures on the same die.

3.2.3 COMPLEMENTARY METAL-OXIDE SEMICONDUCTOR (CMOS)- COMPATIBLE FABRICATION

In terms of fabrication, electron-beam lithography has been used extensively for the fabrication of SOI Bragg gratings. It can make features in the nanometer regime, which is especially important for small corrugations. However, it is unsuitable for commercial applications [17]. Alternatively, we can explore the possibilities of using CMOS fabrication techniques. Deep-ultraviolet (DUV) lithography, especially at 193 nm [11], has been proven to be capable of making high-quality photonic devices in silicon, and, more important, it is CMOS-compatible and can be used for high-volume production. The devices presented in this section were fabricated via ePIX-fab [18] at IMEC using a CMOS-compatible process. The pattern was defined using 193 nm DUV lithography with an ASML PAS5500/1100 step-and-scan system and a dry etching process [19].

3.2.3.1 Fabrication Uniformity

Many silicon photonics devices are highly sensitive to dimensional variations; for example, deviations in width or thickness can cause a spectral response shift for wavelength-selective devices. Although active components (e.g., thermal or electrical tuning) may be required for accurate compensations [20], it is still important to improve the uniformity of passive devices to a practical level.

The results presented in the previous section are measured from devices on a single die. In this section, we will provide more experimental data to show their die-to-die nonuniformity. Here, the mask pattern (including all aforementioned devices) is replicated on a 6-inch multiproject wafer that results in many dies. There are 13 dies in the center row of the wafer, which are labeled –6 to 6 from left to right (die –2 was used for the demonstrations in the previous section). On each die, the grating devices are located within in a small area. For the deep etch process, the exposure dose across the wafer is intentionally increased from left to right for research purposes [11]. This results in a reduction in width for the deep-etched structures (e.g., strip waveguides or slab region of rib waveguides) from die –6 to die 6, and, therefore, is an intentional fabrication variation. On the other hand, the exposure dose for the shallow etch is constant across the wafer. To evaluate the die-to-die nonuniformity, we chose two representative devices: strip gratings (SG) with 40 nm corrugations and rib gratings (RG) with 80 nm corrugations on the slab. Their central wavelength and bandwidth variations are shown in Figure 3.13. It is obvious that both the wavelength and bandwidth variations of SG are much larger than those of RG. For simplicity, we only focus on the wavelength variations here. Based on the simulation results of Wang et al. [8], the wavelength shifts ($\Delta\lambda$) with waveguide thickness and width variations (Δt and Δw) are almost linear, therefore, the slopes can be defined as sensitivities. Table 3.1 lists the simulated sensitivities.

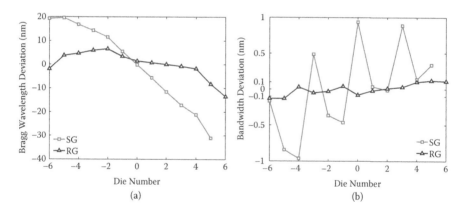

FIGURE 3.13 Measured die-to-die nonuniformity for the two grating devices: (a) Bragg wavelength deviation from mean, (b) bandwidth deviation from mean.

TABLE 3.1
Simulated Bragg Wavelength Sensitivity to Thickness and Width Variations

Structure	$\Delta\lambda/\Delta t$ (nm/nm)	$\Delta\lambda/\Delta w$ (nm/nm)
Strip waveguide gratings (SG)	2.6	1.2
Rib waveguide gratings (RG)	2.5	0.065

Note: For the rib waveguide, the thickness variation is applied to the total thickness and the shallow etch depth is assumed to be constant, and the width variation is only applied to the slab region.

We can see that both structures are more sensitive to thickness variations than to width variations. This is particularly true in the case of the rib waveguide, where the sensitivity to width variations is very low. Again, this is consistent with the earlier conclusion that the optical field intensity in the rib waveguide is very low around the sidewalls.

To extract the sources of the Bragg wavelength variation, we can write a pair of equations describing the variations in a matrix form [8]:

$$
\begin{bmatrix}
\dfrac{d\lambda_{SG}}{dt} & \dfrac{d\lambda_{SG}}{dw} \\[2ex]
\dfrac{d\lambda_{RG}}{dt} & \dfrac{d\lambda_{RG}}{dw}
\end{bmatrix}
\begin{bmatrix}
\Delta t \\[1ex]
\Delta w
\end{bmatrix}
=
\begin{bmatrix}
\Delta\lambda_{SG} \\[1ex]
\Delta\lambda_{RG}
\end{bmatrix}
\tag{3.9}
$$

where λ_{SG} and λ_{RG} are the Bragg wavelength variation of SG and RG, respectively; Δt and Δw are the thickness and width variations; and the terms in the 2×2 matrix are the simulated sensitivities in Table 3.1. By inserting the experimental data in Figure 3.13a into the right side of Equation (3.9), the dimensional variations can then be extracted, as shown in Figure 3.14a. This method is similar to Zortman's work [21], where the TE and transverse-magnetic (TM) modes of a microdisk resonator were used to extract the dimensional deviations. Here, we use two devices instead and made a few approximations to simplify the analysis [8].

From Figure 3.14a, we can see that the width significantly decreases with the die number, which verifies the exposure dose sweep [11]. On the other hand, the thickness variation is random and has a much smaller variation range (the standard deviation is about 2 nm, showing good agreement with previous measurements on similar wafers [19]). However, this does not mean that the thickness variation is less important than the width variation. For the rib waveguide gratings, the thickness variation is actually the dominant source for the Bragg wavelength variation, because the device is much more sensitive to thickness than to width, as shown in

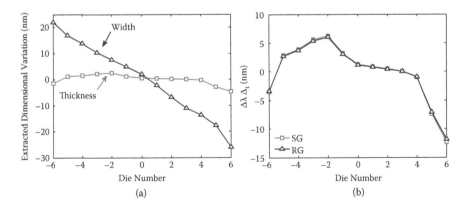

FIGURE 3.14 (a) Extracted dimensional variations based on Equation (3.9). (b) Bragg wavelength shift due to thickness variations.

Table 3.1. Figure 3.14b plots the wavelength variations only due to the random thickness variations, and they are on the order of ±10 nm. This is obviously too much, and it is a critical issue that needs to be addressed before the commercial production of silicon photonics circuits. To improve the SOI thickness uniformity, adaptive process control (e.g., corrective etching) could be used [22].

3.2.3.2 Lithography Simulation

As previously mentioned, silicon photonic Bragg gratings suffer from serious lithographic distortions. Therefore, it is essential to include the effects of the fabrication process in the design flow so that they are properly accounted for [9]. Here, we use an advanced lithography simulation tool [23] to predict the fabrication imperfections for single-etched strip waveguide gratings [10]. First, we chose a device to calibrate the lithography model so that the postlithography simulation fit the experimental data. Then, the model is fixed for all other devices. Here, the device used for calibration is a strip waveguide grating designed with 40 nm square corrugations (named as device A). Figure 3.15 shows the simulation results for device A, and we can see that the corrugations are greatly smoothed. After the lithography simulation, we simulate the spectral responses of the virtually fabricated grating devices using a two-dimensional (2D) finite-difference time-domain (FDTD) method [24], and then compare them with the original design as well as the measurement results of the devices actually fabricated. Figure 3.16b shows the transmission spectra for device A. It can be seen that the original design has a much larger bandwidth than that of the fabricated device. The measured extinction ratio is less than that predicted by the FDTD simulation, and this discrepancy may arise from a number of factors, such as sidewall roughness [25] and measurement limitations. Figure 3.16a shows the results for the device designed with 20 nm square corrugations. Similarly, the postlitho simulation shows a bandwidth that agrees well with the measured bandwidth and is much narrower than that of the original design. Figure 3.17 plots the simulated and measured bandwidths versus the designed corrugation widths. For all the devices,

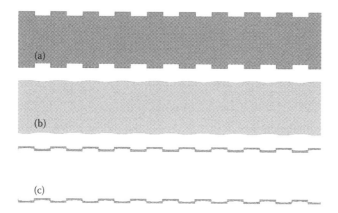

FIGURE 3.15 Lithography simulation for device A. (a) Original design, (b) simulation result, (c) XOR between (a) and (b).

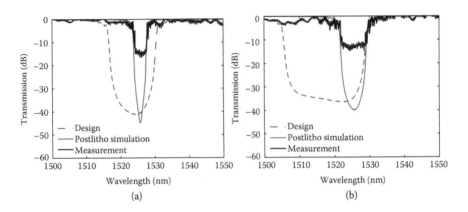

FIGURE 3.16 Comparison for the device designed with (a) 20 nm and (b) 40 nm square corrugations. (After Wang et al., *2012 IEEE Int. Conf. Group IV Photon.*, pp. 288–290, 2012.)

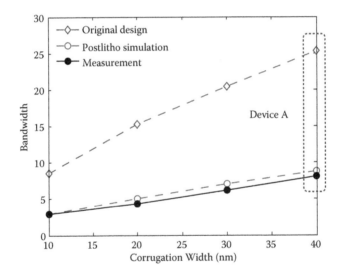

FIGURE 3.17 Bandwidth versus designed corrugation width. (After Wang et al., *2012 IEEE Int. Conf. Group IV Photon.*, pp. 288–290, 2012.)

the postlitho simulation agrees very well with the measurement, whereas the mismatch between the original design and the measurement is very large.

It is worth noting that in fabrication process via optical lithography, it is difficult to optimize the illumination settings for various types of patterns simultaneously, for example, the settings that are optimized for isolated structures such as photonic wires are usually not ideal for dense structures such as photonic crystals [11]. Although lithography simulation can effectively predict the fabrication of waveguide Bragg gratings here, this technique can be applied to many other silicon photonic devices, such as photonic crystals, which are also very sensitive to lithographic distortions.

Lithography simulation is even more necessary at the circuit level or system level. We believe that this is an important step in the direction of design-for-manufacturing in the field of silicon photonics. Of course, there are many issues to be addressed, but fortunately the development of silicon photonic fabrication techniques can benefit greatly from the vast library of knowledge that already exists in the microelectronics industry as well as the continued advancements in this particular field.

3.3 APPLICATIONS OF BRAGG GRATINGS IN SILICON PHOTONICS

3.3.1 NONUNIFORM WAVEGUIDE GRATING STRUCTURES

3.3.1.1 Sampled Gratings

A sampled grating is formed by applying a sampling function to a conventional uniform grating so that the grating elements are removed in a periodic fashion, as shown in Figure 3.18. Since the sampling function can lead to a reflection spectrum with periodic maxima, sampled gratings are often deployed in tunable semiconductor lasers to achieve a wide tuning range through the Vernier effect [26]. Figure 3.19 shows the measured reflection spectrum of a sampled gratings on strip waveguide, where periodic maxima were clearly observed. This comblike reflection spectrum, combined with the potential of using the Vernier effect, makes this device interesting for future applications such as in tunable silicon lasers, multichannel add/drop multiplexers, and dispersion compensations [6].

3.3.1.2 Phase-Shifted Gratings

We know that a uniform grating has a stopband around the Bragg wavelength in the transmission spectrum. If a phase shift is introduced in the middle of the gratings, as illustrated in Figure 3.20, a narrow resonant transmission window will appear within the stopband; therefore it can be used as a band-pass transmission filter [27]. Typically, the length of the phase shift is equal to a half of the grating period so that the resonance peak is at the center of the stopband. Figure 3.21 shows the measured transmission spectra of a fabricated device. It is clearly seen that a narrow resonant peak appears in the center of the stopband. The resonance line width is about 50 pm, corresponding to a high quality factor (Q) of about 30,000.

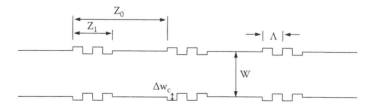

FIGURE 3.18 Sampled grating structure.

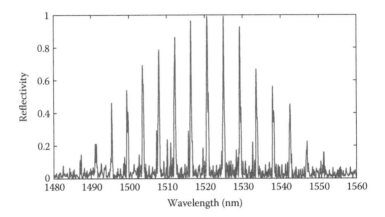

FIGURE 3.19 Measured reflection spectrum of a sampled grating. (More details can be found in Wang et al., *IEEE Photon. Tech. Lett.*, vol. 23, no. 5, pp. 290–292, 2011.)

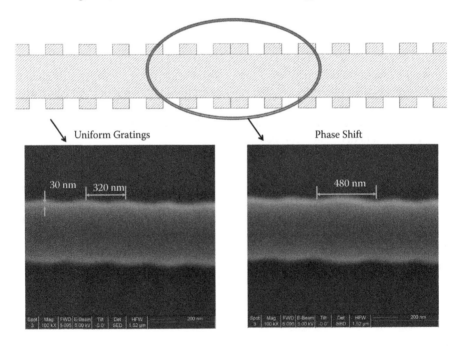

FIGURE 3.20 Phase-shifted gratings. (Top) Schematic diagram. (Bottom) SEM images of a fabricated device. (After Wang et al., *2011 IEEE Photon. Conf.*, pp. 869–870, 2011.)

3.3.2 Grating-Assisted, Contradirectional Couplers

Add-drop filters are essential components for wavelength-division multiplexing (WDM) systems and have been extensively developed for the SOI platform [28]. Among these devices, ring-resonator add-drop filters have received much attention [28,29]. Nevertheless, microring resonators have Lorentzian drop-port responses

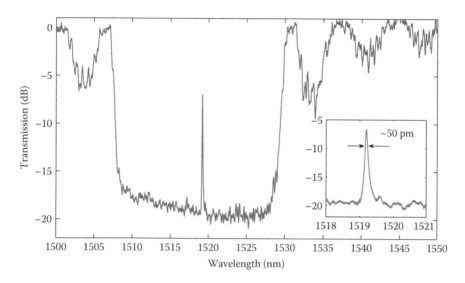

FIGURE 3.21 Measured transmission spectrum of a phase-shifted grating fabricated on strip waveguide. (Inset) Enlarged plot around the resonance peak.

and limited free spectral ranges (FSRs), for which an ideal drop-port response can only be obtained with increased complexity, for example, using series-coupled racetrack resonators with the Vernier effect [30]. Bragg gratings are widely used in optical communications and sensing applications, such as wavelength filters, dispersion engineering, tunable lasers, and reflective sensors. They do not suffer from having an FSR and have a flat-peak spectrum that can be easily tailored by choosing an appropriate dielectric-perturbation structure and geometry parameters, for example, grating period and size, waveguide width, and apodization profile. However, most demonstrated Bragg devices operate in reflection mode (two-port device). This brings about the challenging requirement to integrate an optical circulator.

Grating-assisted, contradirectional couplers (contra-DCs) have no, or very weak, reflection at the operating wavelength and, thus, intrinsically function as wavelength-selective add-drop filters (four-port device), circumventing the need for optical isolators or circulators [31]. Silicon photonic contra-DCs have recently been demonstrated for applications such as WDM filters, dispersion compensation, and nonlinear pulse compression [32–35].

In this section, we will discuss design and properties of silicon photonic contra-DCs. We start with the principle of contra-DCs using coupled-mode analysis. We then discuss waveguide and grating configurations in realizing integrated contra-DCs in the context of CMOS-photonic fabrication. A general procedure of simulation and design considerations will be described. We then discuss a four-port photonic resonator using phase-shifted contra-DCs and the electrical tuning of its response. At the end, we show how to integrate contra-DCs with microring resonators to engineer the responses of microring resonators with wavelength-selective coupling.

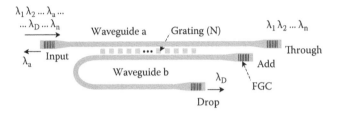

FIGURE 3.22 Schematic of the contradirectional couplers with the fiber grating couplers (FGC) for optical testing. (After Shi et al. *Opt. Lett.*, vol. 36, pp. 3999–4001, 2011.)

3.3.2.1 Principle

As illustrated in Figure 3.22, a contra-DC is a four-port device, consisting of two waveguides with dielectric perturbation formed between the two waveguides. The two waveguides are designed to have significantly different propagation constants and, thus, no or very weak broadband, codirectional coupling exists in the absence of the perturbation. This coupler asymmetry can be easily obtained by varying the waveguide widths due to the high dispersion of SOI waveguides. With the assistance of the perturbation, or grating, band-limited, contradirectional coupling can be obtained near the phase-match condition determined by the perturbation/grating pitch [31].

In this case, the electric field of light in the coupler can be expressed as

$$\mathbf{E}(x,y,z) = A^+(z)e^{-j\beta_a z}\mathbf{E_a}(x,y) + B^-(z)e^{j\beta_b z}\mathbf{E_b}(x,y) \qquad (3.10)$$

where $\mathbf{E_a}$ and $\mathbf{E_b}$ are the normalized electric-field distributions of the transverse modes confined mainly in waveguide **a** and waveguide **b**, respectively, propagating in opposite directions; β_a and β_b are the corresponding propagation constants. The equations governing the coupling between the two modes are [36]

$$\frac{dA^+}{dz} = -i\kappa B^- e^{i\Delta\beta z} \qquad (3.11a)$$

$$\frac{dB^-}{dz} = i\kappa^* A^+ e^{-i\Delta\beta z} \qquad (3.11b)$$

where $\Delta\beta$ is given by

$$\Delta\beta = \beta_a + \beta_b - m\frac{2\pi}{\Lambda} \qquad (3.12)$$

where Λ is the grating pitch and m is chosen to be 1 for the first-order grating design; κ is the distributed coupling coefficient, representing the coupling strength between the two modes, and is given by [36]

$$\kappa = \frac{\omega}{4} \iint E_a^*(x,\,y) \cdot \Delta\varepsilon_1(x,\,y) \, E_b(x,\,y) \, dx \, dy \tag{3.13}$$

where ω is the optical frequency and $\Delta\varepsilon_1$ is the first-order Fourier component of the dielectric perturbation.

Efficient coupling between E_a and E_b only happens at and near the phase-match condition, which is given by

$$\Delta\beta = 0 \tag{3.14}$$

Also, strong intra-waveguide reflections can happen at the Bragg conditions of the individual modes. As an example, Figure 3.23 shows the calculated effective indices of a contra-DCs using a mode solver illustrating how to determine the wavelengths of interest based on the phase-match conditions [36]: $\lambda_a = 2n_a\Lambda$ and $\lambda_b = 2n_b\Lambda$ are the wavelengths for the intrawaveguide Bragg reflections of waveguide a and b, respectively, where n_a and n_b are the effective indices and Λ is the grating pitch; $\lambda_D = 2n_{av}\Lambda$ is the drop-port central wavelength and corresponds to the contradirectional coupling, where $n_{av} = (n_a + n_b)/2$.

Solving Equation (3.11), we can obtain the power coupling efficiency, which is given by

$$\eta = \frac{|\kappa|^2 \sinh^2(sL)}{s^2\cosh^2(sL) + (\Delta\beta/2)^2\sinh^2(sL)} \tag{3.15}$$

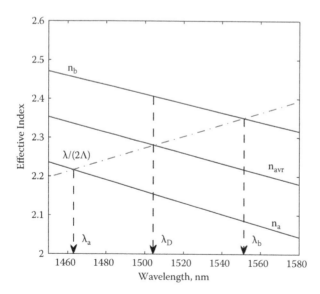

FIGURE 3.23 Calculated effective indices of the first two TE-like modes of a contra-DC using sidewall-modulated strip waveguides.

where $L = N\Lambda$ (N: period number) is the total coupling length and the parameter s is determined by [36]

$$s^2 = |\kappa|^2 - (\Delta\beta/2)^2 \tag{3.16}$$

3.3.2.2 Add-Drop Filters Using Contradirectional Couplers

Here, we show two types of silicon contra-DCs with different waveguide and grating configurations. Their cross-sectional structures are illustrated in Figure 3.24. The first type uses strip waveguides with the grating formed on the waveguide sidewalls. The second type has a rib-waveguide structure with the grating formed on the slab.

The two structures have respective advantages in comparison with each other. Strip waveguides require only one etch and enable a larger spacing between the contradirectional coupling wavelength, λ_D, and the Bragg wavelength, λ_a, due to their stronger waveguide dispersions. However, since light is strongly confined in the strip waveguides, their optical modes are very sensitive to the perturbation. Compared with sidewall-modulated strip waveguides, rib waveguides can have larger corrugations (hundreds of nanometers versus tens of nanometers) due to the lower effective-index contrast. Therefore, they have higher fabrication tolerances and more precise control of weak coupling coefficients needed to obtain narrow-bandwidth filters. As revealed by Equation (3.13), the coupling strength between the two modes is determined by the overlap of the mode profiles and the dielectric perturbation. Therefore, the grating on the slab between the rib waveguides is more efficient since it lies at the intersection of the evanescent waves, as shown in Figure 3.25. In addition, the slab can be doped to form a p-n junction, therefore, the rib-waveguide geometry will be desired for electrical tuning or modulation.

Figures 3.26 and 3.27 show the examples of contra-DCs using strip waveguides [37] and rib waveguides [29], respectively. The devices were designed based on the standard 220 nm SOI wafer and were fabricated by a CMOS-photonic foundry (Imec, Belgium accessed via ePIXfab) using 193 nm lithography. We can see that the contra-DC in strip waveguides has a very wide spacing between λ_a and λ_D of greater

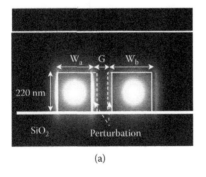

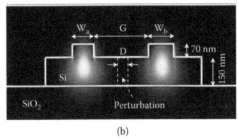

(a)   (b)

FIGURE 3.24 Cross-sectional structures of contra-DCs in (a) sidewall-modulated strip waveguides and (b) slab-modulated rib waveguides, with the coupled modes calculated using a mode solver.

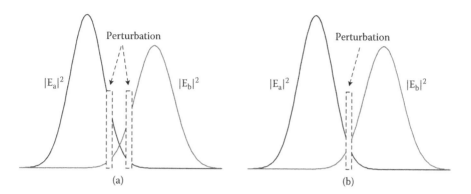

FIGURE 3.25 Two perturbation schemes: (a) sidewall-modulated strip waveguides, (b) slab-modulated rib waveguides.

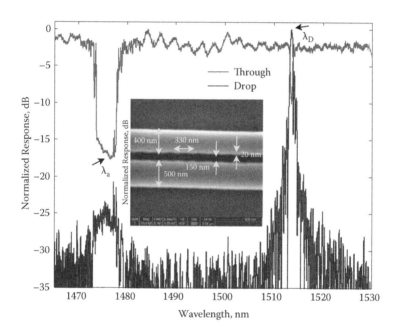

FIGURE 3.26 Through-port and drop-port spectra of a contra-DC in sidewall-modulated strip waveguides. The inset shows an SEM image of the contra-DC. The device has the following parameters: W_a = 400 nm, W_b = 500 nm, Λ = 330 nm, and G = 150 nm, N = 2000. (After Shi et al., "Add-drop filters in silicon grating-assisted asymmetric couplers," *OFC/NFOEC*, OTh3D.3, 2012.)

than 38 nm, able to cover the entire span of the C-band for dense WDM applications; the stopband at λ_D has a 3 dB bandwidth of 0.59 nm and is much smaller than stopband of the Bragg reflection at λ_a, which indicates that the coupling between the forward propagating wave and the backward propagating wave in the input waveguide sees a stronger perturbation than the coupling across the two waveguides.

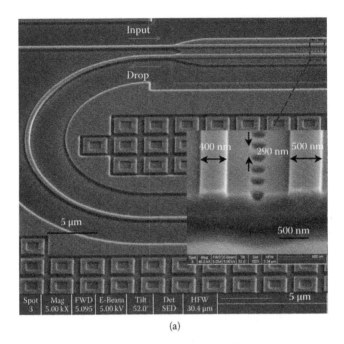

(a)

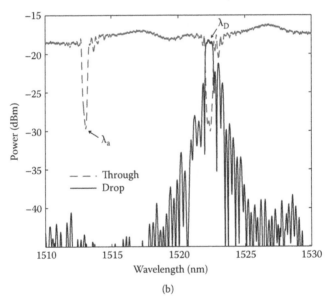

(b)

FIGURE 3.27 A contra-DC in slab-modulated rib waveguides: (a) SEM image; (b) measured spectra (input power of 1 mW with an insertion loss of ~17 dB due to the fiber coupling to the FGCs). The device has a rib height of 70 nm and the following parameters: $W_a = 400$ nm, $W_b = 500$ nm, $\Lambda = 290$ nm, and $N = 4000$, $D = 220$ nm, $G = 1$ μm. (After Shi et al., *Opt. Lett.*, vol. 36, pp. 3999–4001, 2011.)

3.3.2.3 Modeling of Silicon Contradirectional Couplers

Optical design of contradirectional couplers includes the determination of central wavelengths and engineering of response shape and bandwidth. Taking the contra-DC in slab-modulated rib waveguides, shown in Figure 3.27, as an example, a general procedure of simulation using coupled-mode analysis includes:

- Calculate the effective indices of the two coupled modes (first two TE-like modes in this case) as functions of wavelength.
- Find λ_D, λ_a, and λ_b using the phase-match conditions, as shown in Figure 3.28 where the modes are calculated using an eigenmode solver; the spacing between λ_a and λ_D can be controlled by engineering the asymmetry and dispersion of the coupler, for example, varying W_a and W_b.
- Calculate the electric-field distributions of the two modes, as shown in Figure 3.24b.
- Calculate the coupling coefficient, κ, using Equation (3.13).
- Calculate the coupling efficiency, $\eta(\lambda)$, using Equation (3.15).

For reliable design or postprocessing, fabrication errors need to be taken into considerations. For the example shown in Figure 3.27, we notice that the actual corrugation profile is not rectangular, as in the original design, due to the pattern-size effect in the plasma etching. This effect causes weaker coupling strength and thus a narrower bandwidth [6]. A triangular shape is used to approximate the transverse distribution of the dielectric perturbation with a linear transition between the perturbation peak, $\Delta\varepsilon_p$, and the unperturbed section in the longitudinal direction. Then the dielectric perturbation is given by

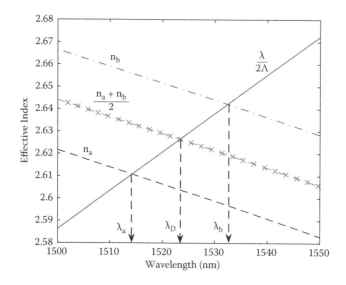

FIGURE 3.28 Calculated effective indices and the phase-match conditions of a contra-DC in slab-modulated rib waveguides.

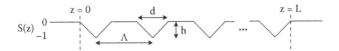

FIGURE 3.29 Longitudinal dielectric perturbation distribution of the device shown in Figure 3.27.

$$\Delta\varepsilon(x,\, y,\, z) = S(z)\Delta\varepsilon_p(x,\, y) \tag{3.17}$$

As shown in Figure 3.29, the periodic function $S(z)$ describes the longitudinal distribution of the perturbation. Now κ can be calculated by

$$\kappa = \frac{\pi c S_1}{2\lambda_D} \iint \mathbf{E}_a^*(x,\, y)\cdot\Delta\varepsilon_p(x,\, y)\, \mathbf{E}_b(x,\, y)\, dx\, dy \tag{3.18}$$

where S_1 is the first-order Fourier-expansion coefficient of $S(z)$. As shown in Figure 3.30, the calculated spectrum is in good agreement with experiment.

For fixed waveguide widths, we can tailor the bandwidth by varying the corrugation width or the coupler gap [32]. The bandwidths measured from the experimental and simulated drop-port responses of the devices with various corrugation widths and coupler gaps are summarized in Figure 3.30b, demonstrating an inverse exponential dependence of the bandwidth on the coupler gap.

3.3.2.4 Electrically Tunable Phase-Shifted Contradirectional Couplers

Silicon photonic resonators are promising for large-scale photonic integrated circuits in a wide range of applications such as optical communications, optical signal processing, and sensing systems [38,39]. A high Q of 100,000 has recently been demonstrated using a transmission filter, or phase-shifted Bragg grating [27]. Like normal Bragg gratings, most Bragg-grating-based resonators operate in reflection mode (two-port device) and have the challenging requirement to integrate an optical circulator. Analogous to the transmission filter, we can introduce a ¼-λ phase shift into the contradirectional coupler to construct a photonic resonator, as illustrated by Figure 3.31. However, different from the phase-shifted Bragg grating discussed in Section 3.3.1.2, this coupled resonator functions as a four-port device.

Figure 3.32 shows cross-section of a design example of an electrically tunable phase-shifted contra-DC [40]. It is based on an SOI rib-waveguide structure with dielectric perturbations formed on both the sidewalls of the waveguide ribs and the slab to achieve a strong coupling coefficient. A ¼-λ phase shift, as shown in Figure 3.31, is introduced in the center of the coupler. As shown in Figure 3.33, there is a resonant peak and, correspondingly, a deep notch in the middle of the through-port spectrum and the drop-port spectrum, respectively, within a stopband of about 7 nm. Although the entire drop-port stopband can be varied by controlling the grating pitch [32], the exact position of the resonant peak is dependent on the length of the phase shift. It is worth pointing out that the filter is resonant at the central wavelength of the drop-port response, that is, at the phase-match condition of the contradirectional coupling that significantly

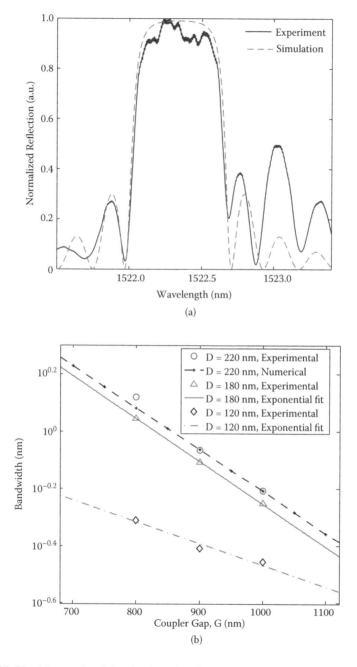

FIGURE 3.30 Measured and simulated results of the contra-DCs using slab-modulated rib waveguides: (a) drop-port spectrum of a device with $[D, G] = [220\ \text{nm}, 1\ \mu\text{m}]$; (b) measured and simulated drop-port bandwidth vs. coupler gap for various sizes of corrugation, showing the inverse exponential relationship. (After Shi et al., *Opt. Lett.*, vol. 36, pp. 3999–4001, 2011.)

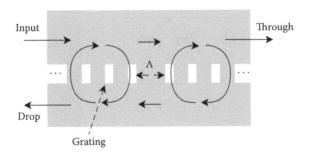

FIGURE 3.31 Schematic top view of a phase-shifted contradirectional coupler.

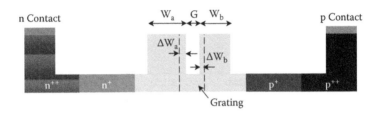

FIGURE 3.32 Schematic top view of a phase-shifted contradirectional coupler. (After Shi et al., *2012 IEEE 9th Conf. Group IV Photon.*, p. WP2, 2012.)

detuned from the Bragg condition of the intrawaveguide reflection, therefore, as opposed to conventional transmission filters using Bragg cavities (e.g., vertical-cavity surface-emitting lasers (VCSELs) or Bragg waveguides [27]), no or very weak Bragg reflection happens at the operating wavelength or within the stopband.

A p-i-n configuration is also implemented in the phase-shifted contra-DC for frequency tuning. The spectrum has a blue shift as the current increases, as shown in Figure 3.34; this is due to the reduced refractive index caused by carrier injection [38]. Noticing that the current is uniformly injected into the whole device along the longitudinal direction, the tuning efficiency may be significantly enhanced by optimizing the overlap of the current density with the longitudinal optical intensity distribution. In this uniform injection case, the electrical tuning shifts both the Bragg stopband as well as the resonant mode simultaneously. For large tuning currents, free-carrier absorption introduces excess loss, thus, reducing the Q and the extinction ratio.

3.3.2.5 Grating-Coupled Microring Resonators

Microring resonators are expected to be essential components in next-generation, integrated photonic circuits. A variety of microring-based devices have been developed for the SOI platform, with applications in optical communications, computing platforms, optical-signal processing, and sensing [29,42,43]. However, they suffer from having limited FSRs, which limits the number of usable channels in WDM systems and detectable range in sensing applications. To selectively excite or suppress longitudinal modes of microring resonators for broader-band operation, major effort has gone into engineering optical ring cavities, for example, using series-coupled

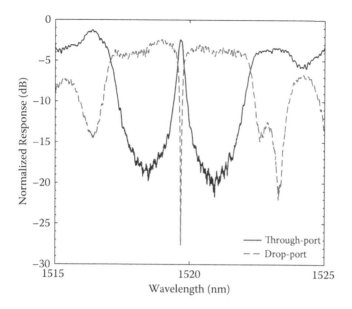

FIGURE 3.33 Measured optical spectra of a phase-shifted contra-DC with the structure illustrated in Figure 3.32. The device has the following design parameters: a rib height of 210 nm; a slab thickness of 110 nm; an input/through waveguide width, W_a, of 600 nm; an add/drop waveguide width, W_b, of 400 nm; a coupler gap, G, of 200 nm; a grating pitch, Λ, of 300 nm; and period number, N, of 700; and a total coupling length of 210 μm. The corrugation amplitudes on the rib sidewalls, ΔW_a and ΔW_b, are 50 nm and 30 nm, respectively. The through-port transmission peak has an out-of-band rejection ratio of more than 17 dB and a 3 dB bandwidth of 0.2 nm corresponding to a Q of about 7,000. The notch in the drop-port response has an extinction ratio of over 24 dB. The device was fabricated by BAE Systems via the OpSIS foundry service. (After Shi et al., *2012 IEEE 9th Conf. Group IV Photon.*, p. WP2, 2012; and Baehr-Jones et al., *Opt. Express*, vol. 20, no. 11, pp. 12014–12020, 2012.)

or cascaded multiple rings with the Vernier effect [30] or inserting Bragg gratings inside the ring cavity [44]. Control of the couplings between the microrings and the bus waveguides is also critical for shaping the spectral responses [45], which is challenging when the bend radius is scaled down to a few micrometers.

Integrating the contra-DC with a microring resonator, selective resonance excitation and thus single-mode operation of the microring are possible [46]. The schematic of a grating-coupled microring resonator is shown in Figure 3.35 where the contra-DC is used as a wavelength-selective coupler. In contrast to broadband directional couplers used in conventional microring resonators, contra-DCs have limited bandwidths that provide microring resonators with an extra degree of freedom for longitudinal-mode selectivity. With one input, optical resonances within the ring cavity could be excited in two directions: clockwise and counterclockwise, associated with the broadband codirectional coupling, κ_b, and the grating-assisted contradirectional coupling, κ_g, respectively. Due to the different propagation constants, the broadband codirectional coupling is very weak, hence the ring oscillates primarily in the direction opposite to a regular ring resonator, that is, counterclockwise for

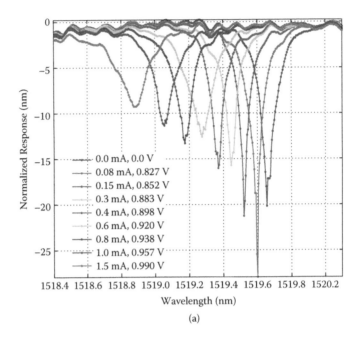

(a)

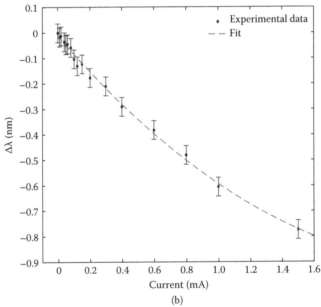

(b)

FIGURE 3.34 Electrical tuning of a phase-shifted contra-DC: (a) drop-port spectra with various currents, (b) resonant-wavelength shift as a function of current. (After Shi et al., *2012 IEEE 9th Conf. Group IV Photon.*, p. WP2, 2012.)

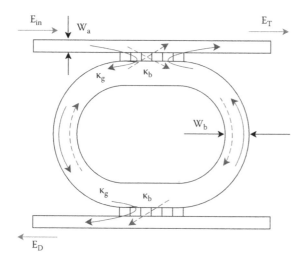

FIGURE 3.35   Schematic of a grating-coupled microring resonator. The solid lines and the dashed lines indicate the optical paths associated with the grating-assisted contradirectional coupling, κ_g, and the broadband codirectional coupling, κ_b, respectively. (After Shi et al., *Appl. Phys. Lett.*, vol. 100, p. 121118, 2012.)

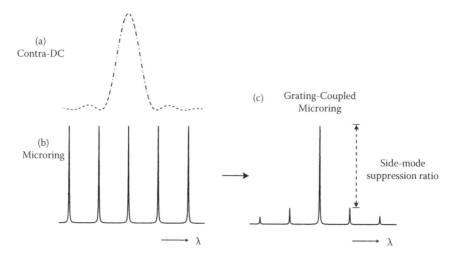

FIGURE 3.36   Schematic of spectral responses of (a) the contra-DC, (b) the microring reso-nator, and (c) the grating-coupled microring resonator. (After Shi et al., *Appl. Phys. Lett.*, vol. 100, p. 121118, 2012.)

a signal at the input port. For the counterclockwise resonance associated with κ_g, efficient coupling between the two waveguides only occurs near the wavelength that satisfies the phase-match condition [31]. Therefore, the longitudinal modes outside of the drop-port passband are not being effectively excited. The operating principle is illustrated in Figure 3.36, showing the drop-port spectrum of the grating-coupled

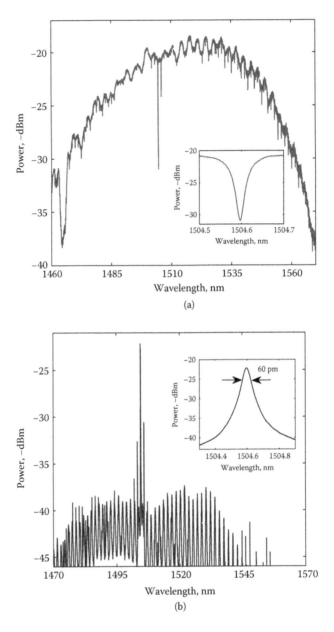

FIGURE 3.37 Spectra of a grating-coupled add-drop microring resonator: (a) through-port; (b) drop-port. The insets show the zoomed-in spectra near the selected longitudinal mode. The through-port spectrum has the envelope of the FGCs' response. (After Shi et al., *Appl. Phys. Lett.*, vol. 100, p. 121118, 2012.)

microring resonator as a result of filtering the spectral response of the microring resonator by the wavelength-selective coupling of the contra-DC.

The spectra of a grating-coupled microring add-drop filter [46] are shown in Figure 3.37. A dominant longitudinal mode is selected within a wide spectral span of over 130 nm. The drop-port spectrum shows a side-mode suppression ratio of more than 8 dB, which is limited by the modes immediately adjacent to the selected mode. Considering the small FSR (1.3 nm) of the ring cavity, this indicates a strong mode selectivity of the contradirectional couplers. The device has an out-of-band-rejection ratio of 19 dB and a 3 dB bandwidth of 60 pm that corresponds to a quality factor (Q) of about 25,000, comparable with microring resonators using broadband directional couplers. This indicates that the gratings do not introduce significant additional losses. In addition to the longitudinal modes of the ring cavity, the through-port spectrum also shows a deep notch at the shorter-wavelength side of the selected mode, which is caused by the intrawaveguide reflection, that is, the Bragg reflection inside the input bus waveguide, as discussed in Section 3.3.2.2. Besides the adjacent modes, the other small resonant peaks are caused by the residual broadband codirectonal coupling, which can be further suppressed by increasing the coupler's asymmetry, that is, the effective-index difference between the coupled modes.

ACKNOWLEDGMENTS

The authors acknowledge contributions from colleagues at the University of British Columbia (Nicolas A. F. Jaeger, Han Yun, Charlie Lin, Samantha Grist, Jonas Flueckiger, Wen Zhang, Mark Greenberg) and colleagues at the University of Washington (Michael Hochberg and Tom Baehr-Jones [now at the University of Delaware], Yang Liu, Li He, Jing Li, Yi Zhang, and Ran Ding). We acknowledge CMC Microsystems for the provision of products and services that facilitated this research, including CAD tools (Mentor Graphics) and design verification (Dr. Dan Deptuck), and fabrication services using imec. Fabrication was conducted at several locations: (1) imec via ePIXfab via CMC Microsystems, (2) BAE Systems via the OpSIS foundry service, and (3) the University of Washington Microfabrication/ Nanotechnology User Facility, a member of the NSF National Nanotechnology Infrastructure Network, by Richard Bojko. We thank Lumerical Solutions Inc. for the design software, particularly Dr. James Pond for useful discussions; Mentor Graphics Corp. for mask design PDK development (Chris Cone) and lithography simulation (Kostas Adam); and funding from the Natural Sciences and Engineering Research Council (NSERC) of Canada and the Canadian Foundation for Innovation (CFI).

REFERENCES

1. J. Buus, M.-C. Amann, and D. J. Blumenthal, *Tunable Laser Diodes and Related Optical Sources,* 2nd ed. John Wiley & Sons, 2005.
2. T. E. Murphy, J. T. Hastings, and H. I. Smith, "Fabrication and characterization of narrow-band Bragg-reflection filters in silicon-on-insulator ridge waveguides," *J. Lightwave Technol.,* vol. 19, no. 12, pp. 1938–1942, 2001.

3. R. Loiacono, G. T. Reed, G. Z. Mashanovich, R. Gwilliam, S. J. Henley, Y. Hu, R. Feldesh, and R. Jones, "Laser erasable implanted gratings for integrated silicon photonics," *Opt. Express*, vol. 19, no. 11, pp. 10728–10734, 2011.

4. D. T. H. Tan, K. Ikeda, R. E. Saperstein, B. Slutsky, and Y. Fainman, "Chip-scale dispersion engineering using chirped vertical gratings," *Opt. Lett.*, vol. 33, no. 24, pp. 3013–3015, 2008.

5. A. S. Jugessur, J. Dou, J. S. Aitchison, R. M. De La Rue, and M. Gnan, "A photonic nano-Bragg grating device integrated with microfluidic channels for bio-sensing applications," *Microelectron. Eng.*, vol. 86, no. 4–6, pp. 1488–1490, 2009.

6. X. Wang, W. Shi, R. Vafaei, N. A. F. Jaeger, and L. Chrostowski, "Uniform and sampled Bragg gratings in SOI strip waveguides with sidewall corrugations," *IEEE Photon. Tech. Lett.*, vol. 23, no. 5, pp. 290–292, 2011.

7. D. T. H. Tan, K. Ikeda, and Y. Fainman, "Cladding-modulated Bragg gratings in silicon waveguides," *Opt. Lett.*, vol. 34, no. 9, pp. 1357–1359, 2009.

8. X. Wang, W. Shi, H. Yun, S. Grist, N. A. F. Jaeger, and L. Chrostowski, "Narrow-band waveguide Bragg gratings on SOI wafers with CMOS-compatible fabrication process," *Opt. Express*, vol. 20, no. 14, pp. 15547–15558, 2012.

9. W. Bogaerts, P. Bradt, L. Vanholme, P. Bienstman, and R. Baets, "Closed-loop modeling of silicon nanophotonics from design to fabrication and back again," *Optical Quantum Electron.*, vol. 40, no. 11–12, pp. 801–811, 2008.

10. X. Wang, W. Shi, M. Hochberg, K. Adam, E. Schelew, J. F. Young, N. A. F. Jaeger, and L. Chrostowski, "Lithography simulation for the fabrication of silicon photonic devices with deep-ultraviolet lithography," *2012 IEEE Int. Conf. Group IV Photon.*, pp. 288–290, 2012.

11. S. K. Selvaraja, P. Jaenen, W. Bogaerts, D. VanThourhout, P. Dumon, and R. Baets, "Fabrication of photonic wire and crystal circuits in silicon-on-insulator using 193-nm optical lithography," *J. Lightwave Technol.*, vol. 27, no. 18, pp. 4076–4083, 2009.

12. I. Giuntoni, D. Stolarek, H. Richter, S. Marschmeyer, J. Bauer, A. Gajda, J. Bruns, B. Tillack, K. Petermann, and L. Zimmermann, "Deep-UV technology for the fabrication of Bragg gratings on SOI rib waveguides," *IEEE Photon. Technol. Lett.*, vol. 21, no. 24, pp. 1894–1896, 2009.

13. I. Giuntoni, A. Gajda, M. Krause, R. Steingrüber, J. Bruns, and K. Petermann, "Tunable Bragg reflectors on silicon-on-insulator rib waveguides," *Opt. Express*, vol. 17, no. 21, pp. 18518–18524, 2009.

14. J. T. Hastings, M. H. Lim, J. G. Goodberlet, and H. I. Smith, "Optical waveguides with apodized sidewall gratings via spatial-phase-locked electron-beam lithography," *J. Vac. Sci. Technol. B*, vol. 20, no. 6, pp. 2753–2757, 2002.

15. G. Jiang, R. Chen, Q. Zhou, J. Yang, M. Wang, and X. Jiang, "Slab-modulated sidewall Bragg gratings in silicon-on-insulator ridge waveguides," *IEEE Photon. Technol. Lett.*, vol. 23, no. 1, pp. 6–9, 2011.

16. R. A. Soref, J. Schmidtchen, and K. Petermann, "Large single-mode rib waveguides in GeSi-Si and Si-on-SiO_2," *IEEE J. Quantum Electron.*, vol. 27, no. 8, pp. 1971–1974, 1991.

17. W. Bogaerts, R. Baets, P. Dumon, V. Wiaux, S. Beckx, D. Taillaert, B. Luyssaert, J. V. Campenhout, P. Bienstman, and D. V. Thourhout, "Nanophotonic waveguides in silicon-on-insulator fabricated with CMOS technology," *J. Lightwave Technol.*, vol. 23, pp. 401–412, 2005.

18. ePIXfab, http://www.epixfab.eu.

19. S. Selvaraja, W. Bogaerts, P. Dumon, D. Van Thourhout, and R. Baets, "Subnanometer linewidth uniformity in silicon nanophotonic waveguide devices using CMOS fabrication technology," *IEEE J. Sel. Top. Quantum Electron.*, vol. 16, no. 1, pp. 316–324, 2010.

20. A. V. Krishnamoorthy, X. Zheng, G. Li, J. Yao, T. Pinguet, A. Mekis, H. Thacker, I. Shubin, Y. Luo, K. Raj, and J. E. Cunningham, "Exploiting CMOS manufacturing to reduce tuning requirements for resonant optical devices," *IEEE Photon. J.*, vol. 3, pp. 567–579, 2011.

21. W. A. Zortman, D. C. Trotter, and M. R. Watts, "Silicon photonics manufacturing," *Optics Express*, vol. 18, pp. 23598–23607, 2010.

22. S. K. Selvaraja, E. Rosseel, L. Fernandez, M. Tabat, W. Bogaerts, J. Hautala, and P. Absil, "SOI thickness uniformity improvement using corrective etching for silicon nano-photonic device," 2011 8th *IEEE Int. Conf. Group IV Photon.*, pp. 71–73, 2011.

23. "Calibre computational lithography," Mentor Graphics, http://www.mentor.com/products/ic-manufacturing/computational-lithography/.

24. "Products," Lumerical, http://www.lumerical.com/tcad-products/.

25. A. D. Simard, N. Ayotte, Y. Painchaud, S. Bedard, and S. LaRochelle, "Impact of sidewall roughness on integrated Bragg gratings," *J. Lightwave Technol.*, vol. 29, no. 24, pp. 3693–3704, 2011.

26. V. Jayaraman, Z.-M. Chuang, and L. A. Coldren, "Theory, design, and performance of extended tuning range semiconductor lasers with sampled gratings," *IEEE J. Quantum Electron.*, vol. 29, no. 6, pp. 1824–1834, 1993.

27. X. Wang, W. Shi, S. Grist, H. Yun, N. A. F. Jaeger, and L. Chrostowski, "Narrow-band transmission filter using phase-shifted Bragg gratings in SOI waveguide," in *2011 IEEE Photon. Conf.*, pp. 869–870, 2011.

28. W. Bogaerts, S. K. Selvaraja, P. Dumon, J. Brouckaert, K. D. Vos, D. V. Thourhout, and R. Baets, "Silicon-on-insulator spectral filters fabricated with CMOS technology," *IEEE J. Sel. Top. Quantum Electron.*, vol. 16, pp. 33–44, 2010.

29. W. Shi, R. Vafaei, M. Á. G. Torres, N. A. F. Jaeger, and L. Chrostowski, "Design and characterization of microring reflectors with a waveguide crossing," *Opt. Lett.*, vol. 35, pp. 2901–2903, 2010.

30. R. Boeck, N. A. F. Jaeger, N. Rouger, and L. Chrostowski, "Series-coupled silicon race-track resonators and the Vernier effect: Theory and measurement," *Opt. Express*, vol. 18, pp. 47–53, 2010.

31. P. Yeh and H. F. Taylor, "Contradirectional frequency-selective couplers for guided-wave optics," *Appl. Opt.*, vol. 19, pp. 2848–2855, 1980.

32. W. Shi, X. Wang, W. Zhang, L. Chrostowski, and N. A. F. Jaeger, "Contradirectional couplers in silicon-on-insulator rib waveguides," *Opt. Lett.*, vol. 36, pp. 3999–4001, 2011.

33. D. T. H. Tan, K. Ikeda, and Y. Fainman, "Chip-scale dispersion engineering using chirped vertical gratings," *Appl. Phys. Lett.*, vol. 95, p. 141109, 2009.

34. D. T. H. Tan, K. Ikeda, S. Zamek, A. Mizrahi, M. P. Nezhad, A. V. Krishnamoorthy, K. Raj, et al., "Wide bandwidth, low loss 1 by 4 wavelength division multiplexer on silicon for optical interconnects," *Opt. Express*, vol. 19, no. 3, pp. 2401–2409, 2011.

35. D. T. H. Tan, P. Sun, and Y. Fainman, "Monolithic nonlinear pulse compressor on a silicon chip," *Nature Commn.*, vol. 1, no. 16, p. 116, 2010.

36. A. Yariv and P. Yeh, *Photonics: Optical Electonics in Modern Communications*, 6th ed. Oxford University Press, 2007.

37. W. Shi, X. Wang, H. Yun, W. Zhang, L. Chrowtowski, and N. A. F. Jaeger, "Add-drop filters in silicon grating-assisted asymmetric couplers," *OFC/NFOEC*, OTh3D.3, 2012.

38. G. T. Reed, G. Mashanovich, F. Y. Gardes, and D. J. Thomson, "Silicon optical modulators," *Nature Photonics*, vol. 4, pp. 518–526, 2010.

39. L. Chrostowski, S. Grist, J. Flueckiger, W. Shi, X. Wang, E. Ouellet, H. Yun, et al., "Silicon photonic resonator sensors and devices," *Proceedings of SPIE*, vol. 8236, p. 823620, 2012.

40. W. Shi, X. Wang, C. Lin, H. Yun, Y. Liu, T. Baehr-Jones, M. Hochberg, N. A. F. Jaeger, and L. Chrostowski, "Electrically tunable resonant filters in phase-shifted contra-directional couplers," *2012 IEEE 9th Conf. Group IV Photon.*, p. WP2, 2012.

41. T. Baehr-Jones, R. Ding, A. Ayazi, T. Pinguet, M. Streshinsky, N. Harris, J. Li, et al., "A 25 gb/s silicon photonics platform," *Opt. Express*, vol. 20, no. 11, pp. 12014–12020, 2012.

42. B. Little, J. Foresi, G. Steinmeyer, E. Thoen, S. Chu, H. Haus, E. Ippen, L. Kimerling, and W. Greene, "Ultra-compact Si-SiO2 microring resonator optical channel dropping filters," *IEEE Photon. Technol. Lett.*, vol. 10, pp. 549–551, April 1998.

43. Q. Xu, B. Schmidt, S. Pradhan, and M. Lipson, "Micrometre-scale silicon electro-optic modulator," *Nature*, vol. 435, pp. 325–327, 2005.

44. A. Arbabi, Y. M. Kang, C.-Y. Lu, E. Chow, and L. L. Goddard, "Realization of a narrow-band single-wavelength microring mirror," *Appl. Phys. Lett.*, vol. 99, p. 091105, 2011.

45. A. Yariv, "Critical coupling and its control in optical waveguide-ring resonator systems," *IEEE Photon. Technol. Lett.*, vol. 14, pp. 483–485, 2002.

46. W. Shi, X. Wang, W. Zhang, H. Yun, C. Lin, L. Chrostowski, and N. A. F. Jaeger, "Grating-coupled silicon microring resonators," *Appl. Phys. Lett.*, vol. 100, p. 121118, 2012.

4 Lasers for Optical Interconnects

Brian Koch

CONTENTS

4.1 INTRODUCTION

Light sources are required for any optical communication system. In most cases lasers are the most practical choice due to their highly directional light output and their tightly controlled and narrow emission wavelengths. These traits make lasers suitable for integration with other components on the same chip and for wavelength multiplexing of different signals together into the same waveguide and fiber. The optimal laser structure and design depends tremendously on the application and the system in which they are meant to operate. For example, depending on the losses in the link and the receiver sensitivity for a given system, the output power required from the laser will vary, and this limits the design space for the laser. Some important considerations in designing a laser structure and its fabrication process typically include cost, size, power dissipation, output power, wavelength properties, and functionality and stability under all possible operating conditions. In this chapter we will discuss semiconductor lasers and their use in optical interconnects. To gain insight into how lasers operate and can be designed, we will begin with some basic laser theory and a discussion of several different types of semiconductor lasers. Then we will discuss general laser design trade-offs and how to approach the optimization of laser designs for different interconnects. In the last section we will discuss some recent experimental demonstrations of semiconductor lasers used in optical interconnects.

4.2 BASIC SEMICONDUCTOR LASER THEORY

In its most general form, a laser consists of an optical gain material placed inside a resonant cavity. The optical gain block in photonics is analogous to an electrical amplifier in electronics, except that optical amplification can occur in more than just one direction. An optical gain material is capable of generating more photons than are sent into it, such that the optical power coming out is higher than that coming in. Obviously this requires some form of input energy, so the gain material must either be optically pumped by light at a different wavelength or electrically pumped with current. This external pumping causes electrons in the gain material being excited to higher energy states. Electrons in excited states can decay to lower energy states by emitting energy in different forms such as heat or light. If there is no light incident on the material (and even if there is incident light), some of the excited electrons will result in spontaneous emission of photons, which will have mainly unpredictable wavelength (within a range determined by the material structure), polarization, and directionality. However, if these excited electrons interact with incident photons, stimulated emission can occur, whereby the excited electrons generate photons identical to incident photons (identical in wavelength, polarization, and direction of propagation).

By placing a gain material inside an optical resonator with feedback, some of the light generated as spontaneous emission can be fed back into the gain material, eventually resulting in stimulated emission and lasing. Because of the properties of stimulated photons mentioned earlier, the light coming out of a laser can be highly coherent. The laser's feedback mechanism can be facilitated by, or may require, some method of guiding the light in a particular direction preferentially (for example, toward the mirrors) such that stimulated emission is more likely to occur and the lasing direction can be controlled. In most semiconductor lasers this function is provided by some form of an optical waveguide. The optical waveguide can often be created in the gain material itself or in another material that is in close contact with the gain material. The principle of an optical waveguide in a laser is very similar to optical waveguides of passive devices or fiber optics, and is based on the use of varying indices of refraction.

An optical resonator is a cavity that typically consists of mirrors or some sort of optical feedback loop. A laser resonator is designed such that some of the light inside the resonator exits (via a partially transmitting mirror or some kind of light splitter), and some of it stays inside providing the feedback. When a gain material is placed inside a resonator with high enough resonance, the optical gain can become equal to the loss through the mirror, and the light power inside the cavity rapidly grows. This is when lasing occurs. Because the feedback occurs only for wavelengths at which the cavity is resonant (i.e., an integer number of wavelengths must match the length of the cavity), only a specific wavelength or wavelengths experience lasing. Here we will present some derivations of simple formulas. More detailed derivations and further explanations can be found in Coldren and Corzine [1] and Agrawal [2].

4.2.1 LASER CAVITY MODES

The wavelengths that are resonant in the cavity and can potentially lase are determined by the cavity length and the index of refraction in the cavity. As shown in Figure 4.1, the electric field in the cavity must have a wavelength such that when the light makes a full round trip in the cavity, it interferes with itself constructively:

$$\lambda = \frac{2n_{\text{eff}}L}{m} \tag{4.1}$$

where n_{eff} is the weighted average effective index of refraction in the cavity, m is the longitudinal mode number, and L is the length of the cavity. This constructive

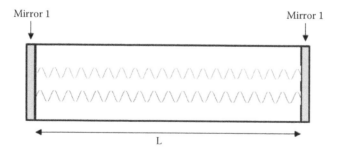

FIGURE 4.1 A simple optical cavity, showing the electric field for two longitudinal modes. Any longitudinal mode of a laser must interfere constructively with itself after a round-trip pass through the cavity.

interference can occur for multiple modes in the cavity, as shown in Figure 4.1 for two wavelengths. The wavelength spacing for adjacent lasing modes is given by

$$d\lambda = \frac{\lambda^2}{2n_g L}$$ (4.2)

where $n_g = n_{eff} - \lambda * dn_{eff}/d\lambda$ is the weighted average group effective index.

4.2.2 THRESHOLD CURRENT AND OUTPUT POWER

In order for lasing to occur, the electric field for the lasing wavelength(s) must also experience zero net loss in a full round trip through the cavity. The electric field in a laser is space and time varying, and can be expressed by

$$\varepsilon = \hat{e} E_0 U(x,y) e^{i\left(\frac{2\pi c}{\lambda}t - \frac{2\pi n_{eff}}{\lambda}z\right)}$$ (4.3)

Here the first term is a unit vector indicating the polarization of the field, E_0 is the magnitude of the field, $U(x,y)$ is the spatial distribution of the mode, t is time, and z is the distance in the direction of propagation. Using Figure 4.2 as a reference, this means that the following equation must be satisfied in order for the electric field to be equal at $z = 0$ and $z = 2L$:

$$r_1 r_2 e^{(\Gamma g_{th} - \alpha_i)L} = 1$$ (4.4)

Γg_{th} is the threshold optical gain of the mode and α_i is the average internal loss in the cavity, as determined by the waveguide loss and other losses in the cavity other than the losses due to light escaping through the mirrors. r_1 and r_2 are the electric field reflectivities of the front and back mirror (which can be replaced by different coupling loss coefficients for ring lasers or other structures). Note that the terms g_{th} and α_i are in terms of optical power and should be divided by 2 in the earlier electric field equation, but since a full round trip in the cavity includes two passes through the gain material, another factor of 2 in front of the length cancels this. This equation shows that the lasing threshold occurs when the gain in the laser becomes equal to

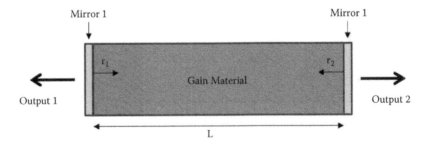

FIGURE 4.2 A simple laser cavity has a gain material placed in a resonant cavity, in this case two mirrors.

all of the loss contributions such as light escaping through the mirrors or scattering in the waveguide. The previous equation can also be expressed with the following equation:

$$\Gamma g_{th} = \alpha_i + \alpha_m \tag{4.5}$$

In this equation, Γ is the confinement factor, which is determined by the fraction of the optical mode that overlaps with the active region in the gain material where carriers are generated. α_m is the mirror loss, given by

$$\frac{1}{L} \ln\left(\frac{1}{R}\right) \tag{4.6}$$

R is the average power reflectivity (electric field reflectivity squared) of the two mirrors.

The laser's threshold current—the injection current at which lasing first occurs instead of just spontaneous emission—is determined by the preceding equations and the relationship between the input current and the gain in the material. Starting with the basic semiconductor active region properties of transparency carrier density (N_{tr}) and material gain coefficient (g_0), the carrier dependent gain (g) in the material is given by:

$$g = g_0 \ln\left(\frac{N}{N_{tr}}\right) \tag{4.7}$$

It is important to note that the transparency carrier density is the carrier density at which the material has no net gain when light propagates through (i.e., the material neither absorbs light nor amplifies it). This is very different than the threshold carrier density of a laser, which is dependent on the mirror loss and other losses. In the Equation (4.7), g_{th} occurs when $N = N_{th}$. To go from carrier density to current, we must consider the volume of the active region (V), the carrier lifetime (τ), and the injection efficiency (n_i), which is the fraction of current injected into the laser's terminals that actually contribute to carriers being generated in the active region:

$$I_{th} = \frac{qVN_{th}}{\tau n_i} \tag{4.8}$$

q is the electron charge. The carrier lifetime is determined by recombination rates occurring in the laser, including spontaneous emission, nonradiative recombination, and carrier leakage.

The output power of the laser (when the current is above the threshold current, I_{th}) is determined from the rate equations and is given by

$$P = n_d \frac{h\nu}{q}(I - I_{th}) \tag{4.9}$$

where h is Planck's constant, ν is the optical frequency, I is the operating current, and the differential quantum efficiency

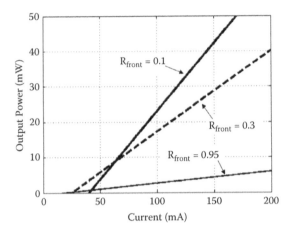

FIGURE 4.3 Light output power versus pumping current for three lasers with the same gain material but different front mirror designs.

$$n_d = \frac{n_i \alpha_m}{\alpha_i + \alpha_m} \tag{4.10}$$

Figure 4.3 shows light output versus pumping current (LI) curves for lasers having the same gain material but different front mirror reflectivities.

The power consumed by the laser is simply given by the operating current multiplied by the operating voltage. The operation current is determined by Equations (4.8), (4.9), and (4.10), and the voltage is determined by the material design, laser structure, and electrical contacts. Lasers that utilize more complex power or wavelength tuning/stabilization circuits will also require additional power. A laser's wall plug efficiency is defined as its output power divided by the total electrical power it uses. This is a common figure of merit for interconnects with minimal power usage requirements.

4.2.3 LASING WAVELENGTHS

We mentioned earlier that multiple longitudinal modes can be emitted from a laser. The lasing wavelength(s) is (are) determined by the combined spectral distributions of the optical gain, the optical loss, the mirror loss, and the cavity mode positions discussed earlier. Figure 4.4 helps explain this concept. The use of wavelength selective mirrors is a common method to obtain single wavelength lasers. This can involve distributed Bragg reflector mirrors, distributed feedback designs, or ring based wavelength filtering. Some small cavity lasers can have very widely spaced cavity modes. In this case single mode operation can be achieved without wavelength selective mirrors if the gain bandwidth of the laser is not wide compared to the cavity mode spacing. We will discuss these concepts more in this chapter when specific types of lasers are addressed.

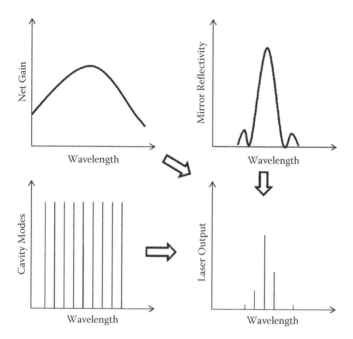

FIGURE 4.4 A laser's output spectrum is determined by the combined effects of multiple elements that make up the cavity.

4.2.4 TEMPERATURE DEPENDENCE AND THERMAL IMPEDANCE

Semiconductor lasers tend to have degraded performance as the temperature in the active region increases. This occurs when the ambient temperature increases or if the laser heats itself. The degradation is a temporary effect caused by factors such as broadening of the material gain peak, changes in the nonradiative recombination rates in the gain material, and increases in the loss due to higher current densities that are required to compensate for the other effects. The gain peak of the material will also shift in wavelength as the temperature changes. This is an important consideration for lasers that have spectral filtering, since the gain peak can become misaligned (or better aligned, depending on the design) with the mirror filter and cause an additional change in the performance. This effect is highly dependent on the specific relationship between the gain peak and mirror reflectivity spectrum, and will not be considered in the following analysis. Considering only lasers that do not have mirrors to provide spectral filtering, the operating current required to maintain a given output power as the temperature changes is given by

$$I = I_0 e^{\frac{T}{T_0}} + I_1 e^{\frac{T}{T_1}} \tag{4.11}$$

I_0 and T_0 describe the effects of the change in threshold, while I_1 and T_1 describe the effects of the change in output power slope. Because the change in threshold current due to temperature has additional effects compared to the change in output power slope, T_1 is usually significantly larger than T_0. Typical values of T_0 are in the range of 50 to 200 K depending on the material system and gain structure, and values of T_1 are generally 2 to 3 times this.

Whereas the ambient temperature rise translates to a nearly identical rise in the active region, the active region temperature can also be changed by the operating current of the laser. The rise in temperature can be simply determined from the laser drive power and the thermal impedance of the laser, Z_T:

$$\Delta T = P_{in} Z_T \tag{4.12}$$

Determining the thermal impedance of the laser is a more difficult problem that depends on the semiconductor material stack, the dimensions of the laser, and eventually how the laser is packaged with a heat sink. These are all very important considerations for lasers that are meant to operate at high temperatures and at high powers.

4.2.5 DIRECT MODULATION

In any optical interconnect the purpose is to transmit data, so the light generated by the laser must somehow have data encoded onto it. This can be done external to the laser using a technique called external modulation, or the laser itself can be modulated using direct modulation. Using direct modulation requires more consideration for the laser than mentioned in the preceding equations. For example, the amount of power required to turn the laser on and off, and the frequency at which this can be done become critical. To fully understand a laser's performance in real applications, it is necessary to consider large signal modulation of the laser's drive current. This requires fairly involved numerical approaches to solving the laser rate equations, which will not be discussed here. However, a common metric that can be used for direction modulation performance is the laser's relaxation resonance frequency. The laser's frequency response has a peak at this value and therefore the higher the relaxation resonance frequency, the higher the frequency that the laser's frequency response will roll off, and the higher the data rate that the laser can transmit. The relaxation resonance frequency can be derived from a small signal analysis of the laser rate equations, and is given by

$$\omega_R = \sqrt{\frac{\Gamma v_g a n_i}{qV}(I - I_{th})} \tag{4.13}$$

Here $a = dg/dN$ is the differential gain as determined by the material, and v_g is the group velocity of light in the laser. As can be seen in the equation, a laser will be better for direct modulation when it has a high differential gain, high confinement factor, high injection efficiency, and small mode volume. Operating at a current that is high above threshold is also beneficial. In practice however, in direct modulation the laser current would need to be modulated between a low current below or very close to threshold and a higher current in order to achieve a reasonable extinction ratio on the modulated signal. Thus achieving high resonance frequency at reasonable modulation powers means that the threshold should be kept as low as possible.

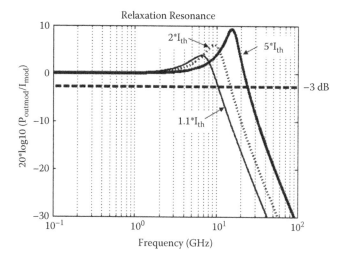

FIGURE 4.5 The relaxation resonance of a laser depends heavily on the bias current. Significant enhancement occurs at higher biases at the expense of drive power.

Figure 4.5 shows the frequency response of a laser of fixed design but different operating currents relative to the threshold. The modulation bandwidth of a laser can be improved by injecting external laser light at one or multiple frequencies offset from the laser's wavelength [3,4]. This creates additional resonances at the frequency corresponding to the wavelength difference of the laser's light and injected light. However, this method obviously requires an additional laser(s) and therefore would likely be reserved only for systems with very high individual channel speed requirements.

Another consideration for directly modulated lasers is the frequency shift or "chirp" that occurs during the modulation. As the carrier density changes with the modulation signal, the lasing wavelength of the laser will shift due to the associated change in the index of refraction. This can lead to signal distortion when propagating along the length of optical fibers or waveguides that have dispersion. In other words, if there is significant chirp and dispersion in the light guide, different parts of the signal travel at slightly different speeds as it propagates leading to problems at the receiver. The frequency chirp is given by the equation

$$\Delta v = \frac{\alpha}{4\pi} \Gamma v_g a \Delta N \tag{4.14}$$

where α is the line width enhancement factor, given by

$$\alpha = -\frac{4\pi}{\lambda a} \frac{dn}{dN} \tag{4.15}$$

In some networks, large amounts of chirp can be tolerated while in others chirp must be very carefully controlled.

4.2.6 LINEWIDTH AND RELATIVE INTENSITY NOISE

Direct modulation leads to a significant broadening of the laser frequency, but even an unmodulated continuous wave (CW) laser has a finite frequency width or linewidth. This factor is again important in links that can have long propagation distances through fibers or waveguides, because dispersion can lead to signal distortion and errors at the receiver. For coherent links this parameter is even more critical. The linewidth of a single mode laser can be expressed as

$$\Delta v = \frac{\left(\Gamma v_g g_{th}\right)^2 n_0}{4\pi P_0} n_{sp} h v (1 + \alpha^2) \tag{4.16}$$

As can be seen from Equation (4.16), it is impacted by many parameters such as the confinement factor, threshold gain, output power (out of the facet of interest, P_0), optical efficiency (out of the facet of interest, $n_0 = n_d/n_i$), the linewidth enhancement factor, and n_{sp}, which is the population inversion factor determined by the energy levels in the gain material.

A related factor for lasers is the relative intensity noise (RIN). Whereas the linewidth is essentially a measure of phase noise, RIN is a measure of the amplitude noise (instability in power). The total RIN noise is a comparison between the laser noise power and the laser's total output power:

$$RIN = \frac{dP(t)^2}{P_0^2} \tag{4.17}$$

RIN can be expressed as a function of frequency or as an average or peak value over all frequencies. Often the RIN is expressed per unit frequency and in decibel scale:

$$RIN_{\frac{dB}{Hz}} = 10 * \log_{10}\left(\frac{RIN}{\Delta f}\right) \tag{4.18}$$

Δf is the frequency range over which the intensity noise was measured. The RIN usually has a maximum at the relaxation resonance frequency of the laser as might be expected. Calculating the expected RIN of a laser can be done using numerical methods based on the rate equations, but it also can be estimated using more basic equations. Even these equations are fairly involved and the form of the equations depends significantly on the operating conditions [1].

Multimode lasers can have a total RIN that is similar to that of a single mode laser. However, typically the power *distribution* is not highly stable. In other words, the power fluctuates for any given mode more than the total power of all modes combined. This is called mode partition noise (MPN). MPN can be a problem for transmission through fiber due to dispersion, as mentioned earlier. Another problem arises when single modes of a multimode laser are isolated and used separately, because the noise for an individual mode can be very large. Even lasers that are "single mode" but have secondary modes that are suppressed by less than 30 dB can suffer significantly from this mode partition noise. These unwanted modes are not necessarily lasing. Typically, suppression of more than 30 dB is necessary to ensure low noise operation, and the amount of noise also depends on the number of secondary modes that exist at the lower power level. These considerations place great importance on the use of adequate spectral mode filtering in lasers designed for certain applications and systems.

Both the RIN and linewidth are determined by the laser itself, but also can be impacted by other elements in a link. In particular, reflections back into the laser can cause surprisingly severe instabilities that directly determine the RIN and linewidth. The effects of feedback also depend on the laser design, but in general it is necessary to keep reflections below –20 dB even for lasers that have a high tolerance. Some laser-based communications systems can be rendered useless even with less than –50 dB reflections. These effects can be calculated using numerical methods involving the laser rate equations, although some more simple equations also exist to estimate the effects [1].

4.3 TYPES OF SEMICONDUCTOR LASERS

4.3.1 FABRY-PEROT LASERS

Fabry-Perot (FP) lasers are one of the simplest forms of semiconductor lasers, as can be seen in Figure 4.6. They consist of a gain material in a waveguide positioned between two mirrors. These mirrors are normally formed by cleaving the material, but can also be formed by etching a facet into the material. Light is emitted evenly from both sides of the laser equally unless one side of the laser is high reflectivity coated. FP lasers are simple and can be inexpensive due to their simplicity, but they typically emit many wavelengths that are not highly controllable due to their lack of wavelength selectivity. These lasers also cannot be integrated with other components if their facets are formed by the edge of a chip. These factors make them less suited to wavelength-division multiplexing (WDM) systems or photonic integration in general. However, due to their simplicity they are a good choice for an inexpensive laser if their performance can be adequate for the intended link.

4.3.2 RING LASERS

Ring lasers are another type of laser that can be relatively simple to fabricate. An example is shown in Figure 4.7. The lasers consist of a ring or racetrack waveguide

FIGURE 4.6 Diagram of a simple Fabry-Perot laser.

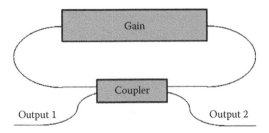

FIGURE 4.7   Diagram of a ring (racetrack) laser.

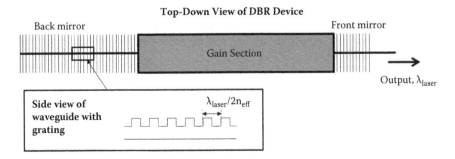

FIGURE 4.8   Diagram of a distributed Bragg reflector laser, with inset showing the side view of the waveguide that has been etched to form a reflecting grating.

that contains a gain section in at least part of its length. The looped waveguide is connected to an output waveguide via a coupler. This coupler can be made from either a multimode interferometer (MMI) or directional coupler. The output power and lasing threshold are determined by how much light is allowed to exit the loop cavity as determined by the coupler design. One issue with ring lasers is that light can propagate in both directions of the ring and therefore light is emitted in two directions simultaneously. In itself this might be acceptable (as in a FP laser), except that in this laser the power in the two directions can be different and changing unpredictably in time. One method to avoid this is to "seed" the lasing in one direction using a low-power light-emitting diode source at one of the output ports, which forces lasing in that direction and emission out one side. This however requires additional power. These lasers are also similar to FP lasers in that they emit light at many different wavelengths unless the cavity has special wavelength selective elements.

4.3.3 DISTRIBUTED BRAGG REFLECTOR (DBR) LASERS

Distributed Bragg reflector (DBR) lasers are lasers that have a gain section positioned between two DBR mirrors. Figure 4.8 is a schematic of a DBR laser. These mirrors consist of a series of small etches into the optical waveguide along the direction of propagation. The spacing of these etches is designed to be ¼ of the wavelength of the laser light (taking into account the index of refraction in the material). Each small etch produces a small reflection, but the combined reflection from all of the etches at the design wavelength is large due to the constructive interference of the multiple reflections at specific wavelengths. In this laser the back mirror can be easily made longer to produce a high reflectivity mirror, while the front mirror can be shorter such that effectively all of the light is emitted from the front. The wavelength selectivity of the laser is determined by the length of the DBR mirrors and the length of the laser cavity. The reflectivity spectrum of such lasers is best simulated using a transfer matrix approach [1], but a closed form equation can also be accurate:

$$r = \frac{\kappa \tanh(\sigma L_g)}{\sigma \left[1 + i \dfrac{\delta}{\sigma} \tanh(\sigma L_g) \right]} \tag{4.19}$$

In this equation $\sigma = \sqrt{\kappa^2 - \delta^2}$ and $\delta = \beta - \beta_0 = 2\pi n_{eff}\left(\dfrac{1}{\lambda} - \dfrac{1}{\lambda_0} \right)$, where λ is the wavelength, λ_0 is the center wavelength of the grating, κ is the grating strength (in 1 cm), and L_g is the length of the grating.

Figure 4.9 shows gratings of different lengths (as might be appropriate for the front and back mirror in a DBR laser) but with the same grating strength. Depending on the cavity mode spacing (the length of the laser cavity), a different reflectivity spectrum will be required to make the laser single mode (if that is desired). The choice of grating strength and length in a laser affect this, and typically it is desirable to design a single mode laser to have more than 30 dB side mode suppression ratio. The calculation of this side mode suppression ratio is too involved for this chapter,

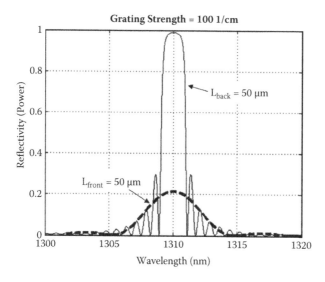

FIGURE 4.9 Typical mirror spectra for a DBR laser's front and back mirror.

but any side modes in the laser certainly need to be suppressed such that their threshold is not reached for any operating current. In practice as the temperature and operating conditions change in a DBR laser, the reflectivity peak and cavity phase will change, resulting in shifts in wavelength and transitions between cavity modes, called mode "hops." These can be compensated over some range by active control of the cavity phase and tuning of the index of refraction of the grating to tune its reflectivity peak, if the proper mechanisms are included in the laser design.

4.3.4 DISTRIBUTED FEEDBACK (DFB) LASERS

Distributed feedback (DFB) lasers are lasers that have the DBR grating in the same region as the gain material, as shown in Figure 4.10. Typically these lasers have a single quarter wavelength shift in the middle of the grating, but designs with longer sections in the middle of the cavity have been investigated, as well as designs with extra shifts in other places along the grating [5,6]. DFB lasers can be considerably shorter than DBR lasers because the mirror occurs in the gain medium, and DFB lasers usually have lower thresholds. Their output powers are usually not as high as DBR or FP lasers due to their short lengths, and their power is emitted from both sides evenly except in special asymmetric designs. Typically this issue is overcome by cleaving at one end of the laser and HR coating the facet to force the light to be emitted almost entirely from the other side. The resulting laser performance, however, depends heavily on the phase of the facet with respect to the gratings and therefore some percentage of the devices become unusable in practice, impacting the yield of these devices. Another option that can be used in some systems is to design a photonic integrated circuit that uses both outputs of a DFB laser. The short length and low threshold of DFB lasers makes them attractive for direct modulation.

4.3.5 VERTICAL-CAVITY SURFACE-EMITTING LASERS (VCSELS)

Vertical-cavity surface-emitting lasers (VCSELs) are similar to DBR lasers in principle except that the cavity is formed in the vertical direction, normal to the surface of the epi material as shown in Figure 4.11. These lasers typically have very highly

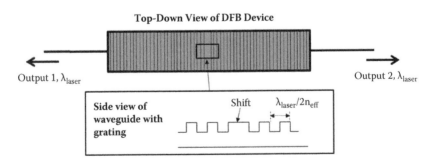

FIGURE 4.10 Diagram of a distributed feedback laser with a ¼ wavelength shift in the middle.

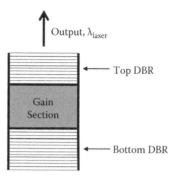

FIGURE 4.11   Diagram of a vertical-cavity surface-emitting laser.

reflecting top and bottom mirrors that are formed by growing different layers of materials with alternating indices of refraction. VCSELs can also have one or both mirrors formed by metal structures and external cavity elements. Reflectivities above 99% are usually desired for both mirrors. This is necessary because the laser cavity is extremely short and thus the gain section is very short, as determined by the thickness of the materials. On the other hand, the confinement factor of a VCSEL can be extremely high and actually can even exceed 1. Threshold currents well below 1 mA are achievable and normally output powers are limited to approximately 0 dBm. Contrary to most in-plane lasers designed for optical interconnects, VCSELs normally have multiple lasing modes and are paired with multimode fibers. While it is possible to make single-mode VCSELs, this requires special processing and impacts their performance [7,8]. Because of their vertical structure, these lasers are more difficult to integrate with other photonic elements, requiring external elements for wavelength division multiplexing or other functions. Fortunately, these lasers are compatible with direct modulation, so integration with external modulators is not necessary. Data rates of 10 Gb/s are standard, with 25 Gb/s appearing feasible for production. Modulation at 40 Gb/s and higher bit rates have also been demonstrated experimentally [9].

4.3.6 WIDELY TUNABLE LASERS

In some systems it is desirable to utilize wavelength tunable lasers. In WDM networks, this can allow for reconfiguration of links and routing of signals, possibly without requiring switches at points between the source and destination. DBR lasers can be tuned to some degree by injecting current into their mirrors to alter the index of refraction and thus the reflected wavelength. This type of tuning is limited to a few nanometers of wavelength range. To achieve wider tuning bandwidths, more complex structures are required.

One example of a widely tunable integrated laser is the sampled grating DBR (SGDBR) laser [10]. This type of laser is similar in structure to a DBR laser, but in place of the front and back DBR mirrors are sampled grating mirrors that have multiple "bursts" of gratings separated by passive sections. Figure 4.12 is a schematic

FIGURE 4.12   Diagram of a sampled grating DBR laser with five bursts of gratings in the back mirror and three bursts of gratings in the front mirror.

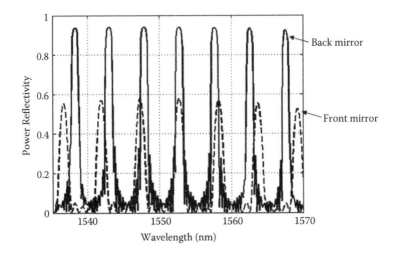

FIGURE 4.13   Typical mirror spectra for a sampled grating DBR laser's front and back mirror. As one or the other mirror is tuned slightly, the wavelength of overlapping reflectivity can change dramatically, causing the laser's wavelength to tune accordingly.

of an SGDBR laser. The front and back mirror are designed with different lengths between the bursts and burst lengths. This design results in the front and back mirror having different reflection spectra with reflection peaks evenly spaced with gaps of wavelengths having no reflection between. When a reflection peak of the front mirror aligns with a peak of the back mirror, lasing will occur at that wavelength. A small amount of tuning in the mirrors can thus result in various overlaps between the front and back mirror reflection spectra across a wide wavelength range. If designed properly, these lasers can use this "Vernier effect" to achieve single longitudinal mode lasing across bandwidths of more than 50 nm. Figure 4.13 shows the reflection spectra of a front and back mirror in a sample grating laser, demonstrating how they overlap with each other.

Similar effects can be achieved using different tuning elements such as super-structure gratings in which the phase of the grating is periodically modulated, or small coupled ring reflectors, or using compound or external cavities. For example, tuning can be done using VCSELs with a microelectromechanical systems (MEMS) tunable external cavity whereby the cavity length is changed enough to significantly alter the wavelength of the lasing mode.

4.3.7 Mode-Locked Lasers

In some systems it is desirable to have sources with pulsed outputs. Pulsed optical sources can relax bandwidth requirements on modulators in systems using the return to zero (RZ) modulation format and can also be useful for systems utilizing time-division multiplexing (TDM). Generating pulses that are shorter than the bit period can be even more demanding than normal direct (or external) modulation because the required frequency response is higher. Mode-locked lasers are a commonly used pulsed source [11]. Figure 4.14 shows the output pulse and associated optical spectrum of an integrated 40 GHz mode-locked laser. These lasers emit multiple longitudinal modes that are locked in phase with each other. Though the electric field of each mode is constant, the interference of the electric fields of these phase-locked modes as they oscillate in the laser cavity results in pulses in the time domain. The phase-locking element inside the cavity can be implemented using various methods.

In integrated semiconductor lasers, the typical phase locking element is a saturable absorber that can be made from the same material as the laser gain material. However, the saturable absorber must be electrically isolated from the gain section so that it absorbs some fraction of the light. The absorber can be reverse biased or left unbiased. The saturable absorber has the characteristic that the amount of light it absorbs decreases as the input power increase, that is, the absorption saturates. When light enters the absorber, the electric fields saturate the material locally. Thus in a laser cavity that contains a saturable absorber, lasing will occur most readily when all electric fields of the different longitudinal modes are aligned in the saturable absorber. In the time domain, the absorber saturates when this pulse begins to pass through and recovers when the pulse leaves the absorber and returns to the gain material, preventing light from forming in the cavity at other times. Figure 4.15 is a schematic of a mode-locked laser, showing the electric fields of four lasing modes in the gain section and below their net optical power from constructive interference in the saturable absorber.

The repetition rate of the pulses emitted from a mode-locked laser is determined by the cavity length and the placement of the saturable absorber in the cavity.

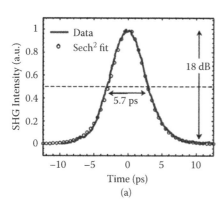
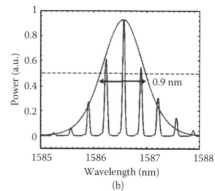

FIGURE 4.14 (a) Output pulse from a second harmonic generation autocorrelator and (b) associated optical spectrum for a 40 GHz integrated mode locked laser.

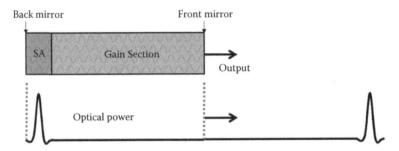

FIGURE 4.15   Diagram of a self colliding-pulse mode-locked laser, showing how the synchronization of electrical fields in the saturable absorber (SA) results in optical intensity pulses forming in the cavity and being emitted at a period corresponding to one round-trip in the cavity.

If the absorber is placed directly next to a mirror in a linear cavity, one pulse forms in the cavity and this pulse collides with itself in the absorber as it reflects. The repetition rate of the pulses is thus determined by the round-trip time for light to pass through the cavity. If the absorber is instead placed in the middle of the cavity (or in any implementation of a ring laser), two pulses actually form in the cavity, traveling in opposite directions. The pulses collide in the saturable absorber, and the repetition rate is twice that of the other case. In some unique examples, the repetition rate can also be altered by purposely filtering specific modes in the cavity. The pulse width is determined by the number of modes that lase, their power distribution, and how well they are locked in phase.

What has been described is a passively mode-locked laser, which generates pulses without any electrical pulses input to it. Mode locking can also be achieved using active locking, which requires a radio frequency (RF) modulation signal applied to the absorber or the gain section (active mode locking can be done without an absorber, but with high power requirements). Hybrid mode locking is a passively mode-locked laser whose performance is enhanced or altered by applying an RF signal to the absorber. In this case the RF power required can be very low (well below 0 dBm) and typically results in lower timing jitter (phase noise) compared to the passive mode-locked laser. It also can allow for more tuning in the repetition rate of the laser.

4.3.8   Multiwavelength Lasers

In WDM systems, sometimes it is desirable to have a laser that simultaneously emits multiple longitudinal modes. This has the potential to be more efficient in terms of power, control electronics, and space compared to an analogous array of separate lasers. As could be inferred from the previous section, mode-locked lasers are one example of a multiple wavelength laser [12,13]. Although the total power coming out of a mode-locked laser oscillates in time, this is due to the interference of the multiple continuous-wave longitudinal modes. When these modes are separated, they each individually have a constant power. An important factor for this type of laser though is the mode competition that results in mode partition noise. Although the noise for the total output power can be low, the noise for an individual filtered mode can be very high. This must be considered carefully for many interconnects.

Other examples of multiple wavelength lasers include arrayed waveguide grating lasers [14], Fabry-Perot lasers, or DBR lasers designed to emit multiple wavelengths. The concern about mode partition noise is true for all of these elements assuming the gain medium is shared between the wavelengths. However, the use of quantum dots as the gain medium instead of quantum wells or bulk material is a way to mitigate this problem [15]. This is because the energy states and thus the different wavelengths are not coupled the same way as in quantum well or bulk-material-based active regions.

4.3.9 OTHER SMALL CAVITY LASERS

Of the laser types we have discussed VCSELs are considerably smaller than all the other types. There are several other types of lasers that have very small cavities and therefore share some similar properties with VCSELs such as low threshold currents, high modulation bandwidth, and to date even lower output powers. These lasers would therefore most likely be used in very short reach applications, possibly limited to within a chip. Microring lasers are small ring lasers that can be coupled to in-plane waveguides [16,17]. These lasers can be combined with other components more readily since they are in-plane structures. Thus they have the potential to be used in integrated WDM transmitters, although controlling their wavelength is not straightforward. Other types of small cavity lasers are photonic crystal lasers [18], and plasmonic or nanolasers [19]. To date all of these lasers are experimental and are not used in commercial links.

4.4 CHOOSING LASER DESIGNS FOR INTERCONNECTS

The general laser traits discussed in Section 4.3 can be used to help determine the best general laser type to use for a given interconnect. The laser characteristics discussed in Section 4.2 (and more characteristics) are also important to consider and will need to be even more carefully analyzed when choosing the specific laser design once the type of laser is chosen. The specific choices will depend on the modulation format (method of encoding data), the link budget, the electrical power budget, and other factors such as manufacturing and engineering development costs. This can become a complex optimization problem because of the interaction terms between the critical parameters, as can be seen by comparing the equations in the previous sections.

4.4.1 CRITICAL FACTORS FOR DIFFERENT LINK TYPES

All of the parameters discussed in the second section of this chapter are important for any laser used in any interconnect. However, certain interconnect types place different restrictions or minimum specifications on these parameters. The best laser for one interconnect application is not necessarily the best for another, even if two links seem similar at first glance. The modulation format is a major factor that will be discussed briefly here. It is important to consider that the laser requirements will ultimately depend heavily on other link factors such as the distance propagated, the type of interconnecting material (type of fiber or waveguide), the other photonic elements between the laser and receiver, and possible environmental effects.

Digital on-off keying (OOK) systems perhaps have the least stringent requirements on the lasers. OOK systems need to have enough power and signal fidelity to be able to distinguish between the 1 and 0 level, and between the subsequent bits. Therefore the critical factors are optical power, intensity noise, and timing jitter at the receiver. Of course the laser has direct impact on these, but every other element in the link also does, and the link in its entirety must be considered.

WDM links based on OOK or other modulation formats require additional considerations. These systems rely on multiplexers and demultiplexers that have certain spectral responses and thus insertion losses and crosstalk levels that depend on the wavelengths of the incoming light. What this means for the laser design is that the wavelength range must be designed to be within this wavelength tolerance of the link. This impacts the laser fabrication methods and the method of control of the lasers' wavelengths. Additionally, depending on wavelength spacing between the channels in the system, the laser may be required to be single mode (the laser may be required to be single mode for other reasons as well). Any time multiple lasers are integrated together, their thermal crosstalk must also be considered. Time-division multiplexing systems, like WDM links, also have additional considerations. For example, they may have more stringent requirements on the timing jitter of the pulses.

Coherent optical communications would have a further requirement of a laser being integrated into the receiver as the local oscillator. These links rely on the interference of the transmitter laser with the local oscillator. This places an additional requirement on the phase noise of each of the lasers being low, meaning that the linewidth must be narrow, typically ~1 MHz or lower. Unless balanced receivers are used in the coherent link, the intensity noise in these links also must be kept significantly lower than in OOK links, meaning the laser RIN must be low. For the differential quadrature phase-shift keying (DQPSK) modulation format, a local oscillator is not required but the source laser linewidth must still be kept low.

4.4.2 CHOICE OF LASER TYPE

The general traits of the various laser types discussed in the previous section can be used as guidelines to determine which types of lasers might be useful in a specific interconnect. Some very basic examples are: a link with single mode fiber cannot use multimode VCSELs, long-distance links must use single wavelength lasers and will likely rely on WDM, links using RZ data or time-division multiplexing may benefit from the use of mode locked lasers, and short-distance links will likely have requirements driven by low cost and high energy efficiency rather than performance. Of course there are many more detailed considerations in choosing a laser type than have been mentioned or could be fit in this chapter.

Another important consideration is the material system and fabrication methods used to make the laser. Conventional systems include GaAs and InP. Lasers made from these materials can be heterogeneously integrated with each other or with other photonic integrated circuit elements on the same chip, can be packaged individually, or can be copackaged with other chips. These types of laser solutions are the most widespread in commercial use today, with varying levels of integration. Recently a significant amount of interest has been directed toward silicon photonics due to the maturity of the processing and potential integration with conventional complementary metal-oxide semiconductor (CMOS) circuitry. However, due to silicon's direct bandgap there is great difficulty in making a silicon-based laser. Significant progress has been made in that effort [20–23], but to date the performance is not adequate for use in practical interconnects. Other methods of utilizing silicon photonics with lasers include novel packaging schemes [24–28] and heterogeneous integration of lasers via wafer bonding [13,29–34].

Choosing between the options for laser design involves considerations that may be less predictable initially, such as the costs of the lasers in manufacturing and their engineering development, and the reliability and expected operating lifetime of the lasers. In practice it may often be necessary to narrow the options without all of the necessary information, by considering only the most critical factors and predicting the trade-offs between the various options.

4.4.3 OPTIMIZATION OF LASER DESIGN

As mentioned earlier, there are many considerations to account for when determining the best laser to use in a link. The optimization problem becomes even harder if there are any unknown variables in the system. Simulations starting from first principles of material design are very difficult, especially if we desire to estimate some of the critical parameters for high-speed modulation and noise phenomena. Instead, or in addition to these material simulations, often basic laser test structures can be measured for a given material design to determine the basic material properties such as gain, loss, and injection efficiency. However, once these test structures are made into real lasers these parameters can change in ways that can be difficult to estimate or simulate. Therefore, it is sometimes necessary to make the full final laser design before it can be fully understood.

Assuming that all of the material parameters are well known, the problem is still fairly complex. As an example we can consider the optimization of the laser length. This simple parameter impacts the laser threshold and output power, relaxation resonance frequency, thermal impedance, linewidth, RIN, tolerance to reflections, the cavity mode spacing and thus the type of mirror that will be required to make the laser single mode, and more. Recall that many of these factors also influence each other. When considering all of the other variables in the laser design (particularly if material design is a variable), it is easy to see how this problem becomes quite complicated. Most factors must be considered simultaneously since they are not truly orthogonal parameters. The importance of certain factors is also not always straightforward. For example RIN from a laser leads to amplitude noise and thus vertical eye closure at the receiver, but in fact also leads to timing jitter and horizontal eye closure when modulation is applied to the CW laser signal. Thus RIN must be accounted for in the power/noise budget and also the jitter budget of a link. It is often a good idea to use simulations to determine an approximate optimal design space and then fabricate and test an array of designs to fully understand the problem and gain full insight into the design trade-offs.

4.5 EXAMPLES OF SEMICONDUCTOR LASERS USED IN OPTICAL INTERCONNECTS

4.5.1 VCSEL-Based Interconnect Examples

Lee et al. [35] demonstrated a 100 meter long, 120 Gb/s link using VCSELs coupled to a multiple-core multimode fiber. The VCSELs were arranged in a two-dimensional array that matched six multimode cores of the multicore fiber. Each VCSEL was directly modulated at 20 Gb/s. At the receiver side, a similarly arranged six-channel surface normal photodetector (PD) array was used. The transmitter used a total power of 550 mW and the receiver used a total energy of 650 mW for a total power dissipation of 10 pJ/bit and a bit error rate below 1e–12. High efficiency links using VCSELs with CMOS driver circuitry and power dissipation as low as 6.5 pJ/bit at 25 Gb/s and 2.6 pJ/bit at 15 Gb/s were demonstrated [36]. VCSELs have also been shown to transmit at data rates of 40 Gb/s, and even 100 Gb/s (with four-level pulse amplitude modulation [PAM]) [9].

4.5.2 Directly Modulated DFB Interconnect Examples

Adachi et al. [37] modulated an array of four directly modulated DFBs at 25 Gb/s each with a bias current of 70 mA and peak-to-peak voltage swing of 0.5 V. The output of these lasers was sent to a 45-degree total-internal-reflection-based mirror that deflected the signals from the in-plane direction to a vertical direction for simpler coupling. These lasers operated up to 85°C and operated in the O-band wavelength range. Sato et al. [38] studied 40 Gb/s direct modulated DFB lasers in the 1.5 micron wavelength range. Bit error rates below 1e–12 were demonstrated with 92 mA bias and 90 mA current swing.

4.5.3 WDM DFBs with External Modulators Interconnect Example

Evans et al. [39] demonstrated a 1.12 Tb/s link using an array of 10 DFB lasers integrated monolithically with Mach-Zehnder modulators on an indium phosphide platform. The signal was transmitted using a polarization multiplexed quadrature phase-shift keying format (PM-QPSK). In this system, two outputs of each DFB laser are modulated separately with different QPSK signals, and each DFB laser has a different wavelength on a 200 GHz grid in the C-band. After modulation the different wavelengths are attenuated if necessary in an array of variable optical attenuators, and then multiplexed on the chip in arrayed waveguide gratings. The two signals that came from the same laser are set to different polarizations and multiplexed together off the chip, after all of the different wavelength signals have been combined for each polarization on the chip. For this modulation format the linewidth of the lasers is of critical importance. Accordingly the linewidths were kept at approximately 1 MHz for all the lasers by optimizing the active material, bias conditions, and by making sure reflections back into the laser were minimized.

4.5.4 EXAMPLE OF SILICON PHOTONICS INTERCONNECT WITH COPACKAGED **DFB** LASER

A 4 × 10 Gb/s transceiver product is described by Dobbelaere et al. [40]. A single DFB laser is sealed in a hermetic silicon micropackage that is epoxy bonded to a CMOS chip. The light is transferred from the laser through an input grating coupler on the CMOS chip, and then split to four paths containing four Mach-Zehnder modulators. At the receiver a polarization splitting grating coupler is used to ensure that light of any polarization is captured. The light from each channel is sent to four high-speed germanium photodiodes connected to on-chip transimpedance amplifiers. A similar 4 × 26 Gb/s transceiver demonstration is also described.

4.5.5 EXAMPLE OF SILICON PHOTONICS INTERCONNECT USING HETEROGENEOUSLY INTEGRATED LASERS

A four channel by 12.5 Gb/s silicon photonics link using coarse wavelength division multiplexing has been demonstrated in references [32–34]. The link's transmitter included an array of four heterogeneously integrated DBR lasers utilizing two different epi materials bonded side by side onto part of the silicon die (two lasers per epi). The lasers operated at 1291, 1311, 1331, and 1351 nm with the wavelengths selected by the silicon DBR mirrors at the front and back of the laser cavity. The outputs of the lasers were sent to an array of four parallel silicon Mach-Zehnder modulators operating at 12.5 Gb/s each, and an Echelle grating to multiplex the signals to a single output waveguide. A silicon/silicon nitride inverted taper was used to expand the optical mode at the output before fiber coupling. The full transmitter was packaged and tested with a silicon photonic receiver containing a demultiplexer and array of four silicon germanium photodetectors. All four channels of the link had bit error rates (BERs) better than 1e–12 at 10 Gb/s when operated simultaneously. At 12.5 Gb/s three of the four channels had BER better than 1e–12 while the fourth was 3e–10.

REFERENCES

1. L. A. Coldren, S. W. Corzine, *Diode Lasers and Photonic Integrated Circuits*, Wiley, New York, 1995.
2. G. P. Agrawal, *Fiber-Optic Communication Systems*, Wiley, New York, 2002.
3. T. B. Simpson, J. M. Liu, A. Gavrielides, "Bandwidth enhancement and broadband noise reduction in injection locked semiconductor lasers," *IEEE Photonics Technology Letters*, vol. 7, no. 7, 1995.
4. J. Wang, M. K. Haldar, L. Li, F. V. C. Mendis, "Enhancement of modulation bandwidth of laser diodes by injection locking," *IEEE Photonics Technology Letters*, vol. 8, no. 1, 1996.
5. G. P. Agrawal, A. H. Bobeck, "Modeling of distributed feedback semiconductor lasers with axially varying parameters," *IEEE Journal of Quantum Electronics*, vol. 24, no. 12, pp. 2407–2414, 1988.

6. S. Ogita, Y. Kotaki, M. Matsuda, Y. Kuwahara, H. Ishikawa, "Long-cavity, multiple phase shift, distributed feedback laser for linewidth narrowing," *Electronics Letters*, vol. 25, no. 10, pp. 629–630, 1989.

7. Y. H. Wang, K. Tai, J. D. Wynn, M. Hong, R. J. Fischer, J. P. Mannarts, A. Y. Cho, "GaAs/AlGaAs multiple quantum well GRIN-SCH vertical cavity surface-emitting laser diodes," *IEEE Photonics Technology Letters*, vol. 2, no. 7, pp. 456–458, 1990.

8. T. -H. Oh, D. L. Huffaker, D. G. Deppe, "Comparison of vertical-cavity surface-emitting lasers with half-wave cavity spacers confined by single- or double-oxide apertures," *IEEE Photonics Technology Letters*, vol. 9, no. 7, pp. 875–877, 1997.

9. R. Rodes et al., "100 Gb/s single VCSEL data transmission link," *Optical Fiber Communication Conference and Exposition (OFC/NFOEC)*, March 4–8, 2012.

10. L. A. Johansson, Y. A. Akulova, G. A. Fish, L. A. Coldren, "Widely tunable EAM integrated SGDBR laser transmitter for analog applications," *IEEE Photonics Technology Letters*, vol. 15, no. 9, pp. 1285–1287, 2003.

11. E. A. Avrutin, J. H. Marsh, E. L. Portnoi, "Monolithic and multi-gigahertz mode locked semiconductor lasers: Constructions, experiments, models, and applications," *IEE Proceedings on Optoelectronics*, 147, no. 4, pp. 251–278, 2000.

12. K. Haneda, M. Yoshida, H. Yokoyama, Y. Ogawa, M. Nakazawa, "Measurements of longitudinal linewidths and relative intensity noise in ultrahigh-speed mode-locked semiconductor lasers," *Electronics and Communications in Japan*, vol. 89, no. 2, pp. 28–36, 2006.

13. B. R. Koch, A. W. Fang, R. Jones, O. Cohen, M. J. Paniccia, D. J. Blumenthal, J. E. Bowers, "Silicon evanescent optical frequency comb generator," *5th IEEE International Conference on Group IV Photonics*, pp. 64–66, 2008.

14. R. Amerfoort, J. B. D. Soole, C. Caneau, H. P. LeBlanc, A. Rajhel, C. Youtsey, I. Adesida, "Compact arrayed waveguide grating multifrequency laser using bulk active material," *Electronics Letters*, vol. 33, no. 25, pp. 2124–2126, 1997.

15. D. Bimberg, N. Kirstaedter, N. N. Ledentsov, Z. I. Alferov, P. S. Kop'ev, V. M. Ustinov, "InGaAs-GaAs quantum-dot lasers," *IEEE Journal of Selected Topics in Quantum Electronics*, vol. 3, no. 2, pp. 196–205, 1997.

16. A. F. J. Levi, R. E. Slusher, S. L. McCall, T. Tanbun-Ek, D. L. Coblentz, S. J. Pearton, "Electrically pumped, room-temperature microdisk semiconductor lasers with submilli-ampere threshold currents," *IEEE Transactions in Electron Devices*, vol. 39, no. 11, 1992.

17. D. Liang, M. Fiorentino, S. Srinivasan, J. E. Bowers, R. G. Beausoleil, "Low-threshold electrically pumped hybrid silicon microring lasers," *IEEE Journal of Selected Topics in Quantum Electronics*, vol. 17, no. 6, pp. 1528–1533, 2011.

18. S. Kita, K. Nozaki, S. Hachuda, H. Watanabe, Y. Saito, S. Otsuka, T. Nakada, Y. Arita, T. Baba, "Photonic crystal point-shift nanolasers with and without nanoslots—Design, fabrication, lasing, and sensing characteristics," *IEEE Journal of Selected Topics in Quantum Electronics*, vol. 17, no. 6, pp. 1632–1647, 2011.

19. K. Ding, Z. Liu, L. Yin, M. T. Hill, J. H. Marell, P. J. van Veldhoven, R. Noetzel, C. Z. Ning, "CW operation of a subwavelength metal-semiconductor nanolaser at record high temperature under electrical injection," *IEEE Winter Topicals*, pp. 15–16, 2011.

20. R. E. Camacho-Aguilera, Y. Cai, N. Patel, J. T. Bessette, M. Romagnoli, L. C. Kimerling, J. Michel, "An electrically pumped germanium laser," *Optics Express*, vol. 20, pp. 11316–11320, 2012.

21. L. Ferraioli, M. Wang, G. Pucker, D. Navarro-Urrios, N. Daldosso, C. Kompocholis, L. Pavesi, "Photoluminescence of silicon nanocrystals in silicon oxide," *Journal of Nanomaterials*, ID 43491, 2007.

22. Y. Gong, S. Ishikawa, S. Cheng, M. Gunji, Y. Nishi, J. Vuckovic, "Photoluminescence from silicon dioxide photonic crystal cavities with embedded silicon nanocrystals," *Physics Review B*, vol. 81, 235317, 2010.

23. S. S. Walavalkar, A. P. Homyk, C. E. Hofmann, M. D. Henry, C. Shin, H. A. Atwater, A. Scherer, "Size tunable visible and near-infrared photoluminescence from vertically etched silicon quantum dots," *Applied Physics Letters*, vol. 98, 153114, 2011.

24. M. Graeme, "Low-cost hybrid photonic integrated circuits using passive alignment techniques," *LEOS 2006*, pp. 98–99, 2006.

25. G. D. Maxwell, "Hybrid integration technology for high-speed optical processing devices," *Optical Internet (COIN)*, pp. 1–2, 2008.

26. G. Maxwell, "Hybrid integration of InP devices," *Conference on Indium Phosphide and Related Materials*, pp. 22–26, 2011.

27. A. Narasimha et al., "A 40-Gb/s QSFP optoelectronic transceiver in a 0.13 µm CMOS silicon-on-insulator technology," *Optical Fiber Communications Conference (OFC)*, 2008.

28. A. Narasimha et al., "An ultra-low power CMOS photonics technology platform for H/S optoelectronic transceivers at less than $1 per Gbps," *Optical Fiber Communications Conference (OFC)*, 2010.

29. A.W. Fang, "Silicon evanescent lasers," Ph.D. dissertation, University of California, Santa Barbara, 2008.

30. A.W. Fang, H. Park, R. Jones, O. Cohen, M. J. Paniccia, J. E. Bowers, "A continuous-wave hybrid AlGaInAs-silicon evanescent laser," *Photonics Technology Letters*, vol. 18, no. 10, pp. 1143–1145, 2006.

31. G. Roelkens, L. Liu, D. Liang, R. Jones, A. Fang, B. Koch, J. Bowers, "III-V/silicon photonics for on-chip and inter-chip optical interconnects," *Laser & Photonics Review*, vol. 4, no. 6, pp. 751–779, 2010.

32. A. Alduino et al., "Demonstration of a high-speed 4-channel integrated silicon photonics WDM link with hybrid silicon lasers," *Integrated Photonics Research Conference (IPR)*, paper PDIWI5, 2010.

33. B. Koch et al., "A 4 × 12.5 Gb/s CWDM Si photonics link using integrated hybrid silicon lasers," *Conference on Lasers and Electro-Optics (CLEO)*, 2011.

34. H. Park, M. N. Sysak, H. Chen, A.W. Fang, Di Liang, L. Liao, B. R. Koch, J. Bovington, Y. Tang, K. Wong, M. Jacob-Mitos, R. Jones, J. E. Bowers, "Device and integration technology for silicon photonic transmitters," *Journal of Select Topics in Quantum Electronics*, vol. 17, no. 3, pp. 671–688, 2011.

35. B. G. Lee et al., "End-to-end multicore multimode fiber optic link operating up to 120 Gb/s," *Journal of Lightwave Technology*, vol. 30, no. 6, pp. 886–892, 2012.

36. C. L. Schow, A. V. Rylyakov, C. Baks, F. E. Doany, J. A. Kash, "25-Gb/s 6.5-pJ/bit 90-nm CMOS-driven multimode optical link," *IEEE Photonics Technology Letters*, vol. 24, no. 10, pp. 824–826, 2012.

37. K. Adachi, K. Shinoda, T. Kitatani, D. Kawamura, T. Sugawara, S. Tsuji, "Uncooled 25-Gb/s operation of a four-wavelength 1.3-µm surface-emitting DFB laser array," *IEEE Photonics Conference (PHO)*, pp. 210–211, 2011.

38. K. Sato, S. Kuwahara, Y. Miyamoto, "Chirp characteristics of 40-gb/s directly modulated distributed-feedback laser diodes," *Journal of Lightwave Technology*, vol. 23, no. 11, pp. 3790–3797, 2005.

39. P. Evans et al., "1.12 Tb/s superchannel coherent PM-QPSK InP transmitter photonic integrated circuit (PIC)," *Optics Express*, vol. 19, B154–B158, 2011.

40. P. Dobbelaere et al., "Silicon photonics for high data rate optical interconnect," *IEEE Optical Interconnects Conference*, pp. 113–114, 2012.

5 Vertical-Cavity Surface-Emitting Lasers for Interconnects

Werner H. E. Hofmann

CONTENTS

5.1 THE APPLICATION OF VERTICAL-CAVITY SURFACE-EMITTING LASER (VCSELS) IN INTERCONNECTS

This section clarifies that VCSEL-based optical interconnects are going to replace the conventional copper-based technology. VCSELs can deliver higher bandwidths below the cost of copper and enable highly scalable solutions with their small footprint. Future supercomputers can only be realized by the broad application of VCSELs.

The maximum data rate an interconnect can carry is limited by Shannon's law. According to that, bandwidth or the signal-to-noise ratio has to be increased for ultimate data rates. As copper suffers from higher damping and crosstalk at higher frequencies, there are physical restrictions in interconnect speeds based on copper. Of course, one can push out the border by higher signal levels and multiple parallel lines. This has happened in the past with CPUs growing more pins and carrying a bigger heat sink on top.

Microelectronics can deliver much more compact circuits than integrated optics due to the De Broglie wavelength of electrons being much shorter. Taking the unimaginable amounts that have gone into the development of silicon microelectronics into account, it is unrealistic that microelectronics will be abolished soon. On the other hand, assuming a certain bandwidth-length product, optics is more energy efficient for interconnects.

Even though optical interconnects are the way to go, the world has tried to stay with copper for interconnects, postponing the inevitable transition to optics for many times. One reason might be that optical communications technologies were almost exclusively driven by the long-haul market. Even though the technological achievements were remarkable, scaling down optical long-haul equipment in price and energy consumption is not straightforward, if not impossible. Consequently, novel technologies had to be developed to meet the requirements of short optical interconnects.

VCSELs are an answer to the question that optical technology could be a workhorse of future interconnects. VCSELs are capable of delivering highest modulation speeds beyond 40 Gbps at high operation temperatures [1]. At the same time they consume orders of magnitude less power and can be mass fabricated at very low cost.

5.1.1 The End of Copper-Based Interconnects

With optical interconnects developed for long-haul applications being not cost competitive to copper, electrical interconnects were squeezed to their limits.

Today's silicon chips are limited by their thermal budget. Surprisingly, most of the heat generated comes from the signal and clock lines [2]. A transition from complementary metal-oxide semiconductor (CMOS)-compatible self-passivating aluminum lines to encapsulated copper has already been realized some years ago. This great technological effort was motivated only by a quite small difference in conductivity. This example shows how uncompromising roadmaps and scaling rules are.

Each new generation of supercomputers, to give another example, is required to deliver a vast boost in computational speed but has to keep cost and energy consumption on a moderate level. This is no longer possible with copper-based interconnects for long. This is why already in 2008 IBM's PetaFlop Supercomputer incorporated 48,000 optical links. Following the roadmap dictated by the market, 2020's ExaFlop mainframes will need 320 million optical interconnects running at 5 times the speed at 1/50 of the energy consumption and 1/400 of the price compared to the optical links used in 2008 [3]. One should also note, that most of the energy consumed by these computers is needed for the interconnects.

Supercomputers used optical interconnects at a quite early stage for two reasons. First, the bandwidth requirements of these high-performance systems could only be accommodated by optics, and, second, the tolerance of higher cost. Other systems would already have greatly benefited from optical interconnects as well, but the cost constraints were too strong inhibiting the widespread use of optical interconnects in the past.

5.1.2   VCSELs for Optical Interconnects

The easiest high-speed optical source is a directly modulated laser. If lasers have to provide excellent beam quality at low cost and energy consumption, VCSELs are the current technology of choice. This is, for example, the reason why each laser mouse for the personal computer is a VCSEL mouse [4].

In order to be applied in short optical interconnects, VCSELs have to deliver high serial bandwidths, with small footprints allowing dense packaging and uncooled operation.

The serial bandwidth is given by system design rules like Amdahl's law, stating that any data processing installation needs to provide certain interconnect bandwidth and memory capacity matching its computational power to avoid bottlenecks.

On the other hand, the amount of optical links ramps up complexity and cost. Moreover, the scalability of a certain technology is also limited by the amount of links that can be connected. System designers from Google stated in 2011 that 40 Gbps would be the bandwidth desired for its next generation of data centers [5].

For system scalability, very dense packaging of the optical chips is a necessity. The VCSEL footprint, being one order of magnitude smaller than edge emitters, is also in favor of this technology. Furthermore, to ensure a compact hybrid package, bottom-emitting devices with the electrical fan-out on the one side and the optical on the other have been proposed by IBM's TERABUS project [6]. This requires the substrate to be transparent requiring longer wavelengths like 980 nm and cannot be accommodated by the 850 nm VCSEL standard designed for multimode fiber links targeting at some 100 m link lengths. Additionally, 980 nm VCSEL devices allow the use of binary distributed Bragg mirrors with much better thermal conductivity than their ternary counterparts. This enables these devices to operate at much higher ambient temperatures that occur in uncooled dense arrays or when integrated on top of a high-performance silicon CPU or memory. Last, but not least, VCSELs have a much smaller energy consumption than other types of laser diodes and would just for that reason be the light source of choice in future application having limited natural resources in mind [7].

As the optical interconnect market requires a volume of billions of devices to be available, the mature GaAs-based technology seems to be the workhorse for the years to come. On the other hand, for integrated optoelectronics on silicon, longer wavelengths like 1.3 or 1.55 μm are very attractive for seamless solutions. Data could be transmitted throughout fibers directly into the silicon waveguides of integrated chips. So longer wavelength VCSELs might be the next step of evolution [8].

Further shrinking of footprint and energy consumption, together with higher bandwidth are future requirements. Novel technologies like nanolasers might replace the newly evolving VCSEL technology one day [9].

5.2   SPEED LIMITATIONS IN VCSELS

As the system cost scales with the amount of interconnects, high serial bandwidths are desired to master the growing data traffic at low cost. Directly modulated VCSELs have bandwidth limitations that have to be conquered to enable highest bandwidths at ultralow cost.

The continuously rising bandwidth demand requires faster lasers. Nowadays, novel communication standards, like 100G Ethernet have to be fixed without having devices ready mastering such kind of modulation speed. Therefore, it is utmost important to understand the mechanisms in a laser under modulation, in order to identify the limiting factors and ultimately minimize them.

One of the main advantages of semiconductor lasers is that they can be directly modulated at high speeds, that is, the direct conversion of an electrical input signal into an optical signal. Directly modulated lasers share the same circuit for biasing and modulation, and no external modulator is needed. Therefore, more cost-effective optical communications solutions, especially interconnects, are possible. In contrast to that, DFB lasers with external Mach-Zehnder modulators are state of the art for long-haul transmission. But already for metro-range links and fiber to the home (FTTH) solutions, and optical interconnects, directly modulated VCSELs are favored as they can provide broadband data links at much lower system cost [10]. Furthermore, VCSELs have a much lower energy consumption and footprint compared to edge emitters, enabling small form-factor communication modules with low power budgets. This point becomes more important as power consumption and heat dissipation is now becoming the limiting factor for data centers and central offices [11,12]. "Green IT" is an upcoming issue, demanding novel technologies like VCSEL-based optical interconnects.

For these applications in digital communication, the maximum speed at which information can be transmitted is depending on the laser's modulation bandwidth. Additionally, the transfer function of the laser should be free of peaking or resonances within the used frequency range. VCSELs, with their high intrinsic damping, are superior in this aspect compared to edge emitters. This is why VCSELs usually allow quite high data rates compared to their bandwidth.

5.2.1 HIGH-SPEED VCSELs

To identify the limits of modulation performance, a rate-equation analysis is a useful instrument [13,14]. For directly modulated lasers the easiest model assumes a photon and a carrier reservoir that interact by stimulated emission.

The rates of change in carrier and photon densities follow these expressions:

$$\frac{dN}{dt} = J_{inj} - J_{th} - R_{stim}S \tag{5.1}$$

$$\frac{dS}{dt} = S\left(\Gamma R_{stim} - v_g(\alpha_i + \alpha_m)\right) + \Gamma J_{sp} \tag{5.2}$$

with N as carrier density and S as photon density in the active region and the optical cavity, respectively. As one has to look at the particles flowing per unit time, the particle numbers have to be scaled by the volumes of the active region and the photon reservoir. Here, this is taken into account by the confinement factor Γ. The first equation states that the carrier density change equals the injected carrier density,

J_{inj}, minus the carriers recombining due to spontaneous emission or losses (J_{th}) and stimulated emission ($R_{stim}S$), whereby R_{stim} is the stimulated emission rate.

For investigations of the dynamic response of the laser, we must analyze these rate equations with the time derivates included. Unfortunately, an exact analytical solution of the full rate equations cannot be obtained. Linearizing the system matrix for small signals, the determinant yields a well-known 2-pole filter function.

The system matrix becomes

$$\begin{pmatrix} j\omega + \mu_{11} & \mu_{12} \\ -\mu_{21} & j\omega + \mu_{22} \end{pmatrix} \begin{pmatrix} dN \\ dS \end{pmatrix} = \begin{pmatrix} dJ_{inj} \\ 0 \end{pmatrix} \equiv \mathbf{Mx} = \mathbf{i} \tag{5.3}$$

with

$$\mu_{11} = \frac{\delta}{\delta N} J_{th} + v_g aS \cong v_g aS \tag{5.4}$$

$$\mu_{12} = v_g g - v_g a_p S \cong v_g g_{th} \tag{5.5}$$

$$\mu_{21} = \Gamma \frac{\delta}{\delta N} J_{sp} + \Gamma v_g aS \cong \Gamma v_g aS \tag{5.6}$$

$$\mu_{22} = -\Gamma v_g g + v_g(\alpha_i + \alpha_m) + \Gamma v_g a_p S = J_{sp} \qquad \Gamma/S + \Gamma v_g a_p S \cong 0 \tag{5.7}$$

The assumptions in Equations (5.4) through (5.7) can be made above threshold neglecting gain compression.

Solving the Equation (5.3) we yield

$$\mathbf{x} = \mathbf{M}^{-1}\mathbf{i} = \frac{dJ_{inj}}{det\mathbf{M}} \begin{pmatrix} j\omega + \mu_{22} \\ \mu_{21} \end{pmatrix} = \begin{pmatrix} dN \\ dS \end{pmatrix} \tag{5.8}$$

and

$$det\mathbf{M} = \omega_R^2 + j\omega\gamma - \omega^2 \tag{5.9}$$

with

$$\omega_R^2 = \mu_{11}\mu_{22} + \mu_{12}\mu_{21} \text{ and } \gamma = \mu_{11} + \mu_{22} \tag{5.10}$$

to be identified as relaxation oscillation frequency ω_R and damping factor γ.

As these equations include many approximations and assumptions we should state that numerical methods would give better results. On the other hand, the non-linearity and uncertainties in real devices are so large that understandable and intuitive formulas are still very helpful.

Damping and resonance are coupled by the K-factor giving the limits for overdamping. For easy modeling, parasitics can be added by first-order low-pass filter. In various textbooks we can find further calculations finding limits caused by parasitic, device heating, damping, and so on. Unfortunately, the assumptions made to keep these formulas easy might be valid for edge-emitting lasers operating around 2.5 Gb/s, however, they do not hold for VCSELs beyond 10 Gb/s. The real limits for high-speed VCSELs are:

- Parasitics that are directly limiting the over-damped responses
- Damping, which gives a real physical limit and increases the effect of parasitics
- Limitations of the relaxation oscillation frequency, f_R, by device heating, the active medium, and the laser cavity
- Other nonlinear effects like transport, spatial hole burning, current crowding, modal properties, and so on.

To identify intrinsic limitations, we have to look at the original equations gained by the small-signal rate-equation analysis:

$$4\pi^2 \cdot f_R^2 = v_g^2 \cdot g_{th} \cdot \Gamma \cdot a \cdot S \tag{5.11}$$

$$K \equiv \frac{\gamma - \gamma_0}{f_R^2} \approx \gamma / f_R^2 = \frac{4\pi^2}{v_g} \cdot \frac{1}{g_{th}} \left(\frac{1}{\Gamma} + \frac{a_p}{a} \right) \tag{5.12}$$

The resonance frequency (f_R) as given in Equation (5.11), is limited by group velocity (v_g), threshold gain (g_{th}), the confinement (Γ) of the optical wave, the differential gain (a), and the photon density (S). For high modulation speed, a high resonance frequency is preferential. By raising the resonance frequency, also the damping rises causing a limitation of the performance. The figure of merit to judge how fast damping rises with higher resonance frequencies is the K-factor, which is also connected to device parameters. The interrelationship is given in Equation (5.12) with a_p as the gain compression factor. Even though the damping flattens out the response, which can be beneficial in some special cases [15], small K-factors are preferential for ultimate modulation speeds. Consequently, threshold gain, optical confinement, and differential gain have to be maximized. Additionally, high photon densities at low gain compression are needed. Last but not least a very good thermal design is crucial, as otherwise a high photon density will never be achieved and all other figures of merit will degrade. Generally speaking, high-speed performance always requires very good static laser performance as a precondition.

Even though these considerations are generally independent of the laser wavelength, the desired emission wavelength also has an impact on the potential device speed. This is caused by two reasons. First, longer-wavelength devices require lower-bandgap materials, which provide inferior carrier confinement. Additionally nonradiative effects like Auger are more pronounced and free-carrier absorption rises by the power of two with wavelength [16].

On the other hand, there are only a couple of device technologies available, for example, devices grown on GaAs or InP substrates. Each of these technologies has

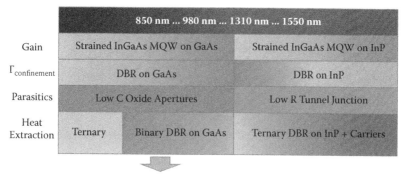

980 nm VCSEL for High Speed

FIGURE 5.1   The impact of emission-wavelength dependent device technology on potential modulation speed. (See References)

some intrinsic limitations based on the desired emission wavelength. These facts are schematically depicted in Figure 5.1, giving an overview of certain laser wavelengths favored by nature assuming state-of-the-art device technology. It turns out that emission wavelengths around 1 µm seem to be a very promising candidate for directly modulated laser sources operating at ultimate speeds [18].

5.3   HIGH-SPEED, TEMPERATURE-STABLE 980-NM VCSELS

The recent development of ultra-high speed, temperature-stable 980 nm VCSEL devices is described in this section. Data rates in excess of 40 Gb/s and data transmission at temperatures as high as 155 °C are demonstrated.

Compared to 850 nm, the wavelength of 980 nm has several advantages for use in short distance optical interconnects. The transparency of gallium arsenide at this wavelength allows realizing of bottom-emitting devices, which enable high packaging density and simple integration to silicon photonics. Additionally, binary GaAs/AlAs-distributed Bragg reflectors (DBR) can be used. By replacing the commonly used ternary alloys with binary material, the thermal conductivity of the mirrors can be enhanced significantly leading to effective heat extraction from the VCSELs. This is essential for highly temperature-stable, high-speed performance. Temperature stability can contribute to low power consumption of optical interconnects, because high-speed operation at constant current and voltage driving parameters provides the opportunity to dispose of cooling systems and to use simpler driver feedback circuits [17, 19].

The high-speed, temperature-stable 980 nm VCSELs made at the Center of Nanophotonics at the Technical University of Berlin [1] were processed in a coplanar low-parasitic ground-signal-ground contact pad layout incorporating thick benzocyclobutene (BCB) passivation layers as depicted in Figure 5.2a. Moreover, we widely varied the device layout to confirm the best device geometries as presented in Figure 5.2b. To identify the best design we used our homebuilt, fully automated wafer prober to do the mapping of various device characteristics of a whole 3-inch wafer. This allowed us to identify the best device design for various purposes. Consequently, we fabricated VCSELs with a wide range of oxide-aperture

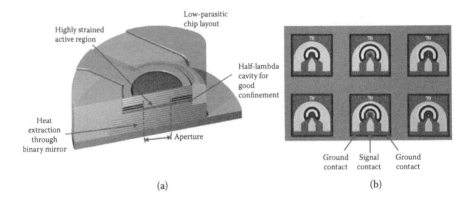

FIGURE 5.2   980-nm high-speed VCSEL. (a) Schematic of a VCSEL chip in a high-speed layout. (b) Photographic picture of VCSEL-chips on wafer showing systematic design variation. The chips can be directly measured using high-frequency probes in a ground-signal-ground layout. (See References)

diameters from 1 to 10 μm. These VCSELs were built upon the previously reported oxide-confined 980 nm devices [20]. To achieve higher speed we reduced the cavity length to λ/2. GaAs/AlAs mirror pairs are used in the n-doped bottom DBR not only to reduce penetration and effective cavity length but also to increase heat extraction from the active region. The mirror reflectivity of the out-coupling mirror was reduced to lower the photon lifetime or threshold gain, respectively, and achieve highest modulation speed. Multiple oxide apertures were used to reduce the parasitic device capacitance. Furthermore, the apertures were moved very close to the active region to avoid damping of the resonance-frequency response by carrier transport. The apertures were formed by an in situ controlled wet oxidation process utilizing our homebuilt oxidation furnace with precise vapor pressure, temperature, and flux control. To avoid spatial hole burning we replaced the phosphorous-containing barriers by pure GaAs with higher electron mobility and better thermal conductivity. The graded interface around the active region was replaced by an abrupt interface for better carrier confinement at high operation temperatures. The mode-gain offset was 15 nm for optimal room-temperature performance. Furthermore, the highly strained InGaAs multiple quantum wells (MQW) were tailored to achieve maximum photoluminescence without gain broadening. From this measure better gain properties are expected, being advantageous for the overall device performance. To avoid carrier reservoirs and storage for hot carriers, any tapered border layers around the active region were eliminated. Sometimes these tapers are named "separate confinement heterostructure" layers (SCH) as the epi-design looks very similar to the successful SCH-confinement of edge-emitting lasers.

To sum up, the high-speed, temperature-stable 980 nm VCSELs made at the Center of Nanophotonics at the Technical University of Berlin feature:

- Low parasitic chip design
- Multiple oxide apertures (low capacitance)

- Sophisticated modulation doping (low resistance, low optical loss)
- Optimized geometrical chip and contact pad layout
- High-speed active region
- Highly strained InGaAs MQWs for a large differential gain
- GaAs barriers for high electrical and thermal homogeneity to avoid gain compression
- Excellent optical confinement
- Half-lambda cavity
- Short DBR penetration due to the high index-contrast binary bottom mirror
- Minimized transport effects
- No "separate confinement heterostructure"-like gradings next to the active region
- Oxide aperture next to the active region avoiding current crowding
- Highly optimized thermal design
- Heat extraction via binary GaAs/AlAs mirror supported by a double-mesa chip
- Active region placed very close to heat sink
- Strong carrier confinement by $Al_{0.90}Ga_{0.10}As$ and the elimination of carrier reservoirs

5.3.1 HIGH-SPEED VCSEL MODULATION

In order to evaluate the high-speed performance of a fiber-based optical system, both sender and receiver must be able to provide enough bandwidth. As the 980-nm waveband is rather new for optical communication systems, and ultrahigh-speed lasers were not available until recently, there are no proper receiver modules on the market yet. Therefore an optical receiver module within a collaborative research project capable to receive data rates in excess of 40 Gb/s also had to be developed. The experiments with data rates beyond 25 Gb/s were carried out using a demonstrator from u²t with ~30 GHz bandwidth, multimode fiber input, a responsivity of ~0.26 A/W @980 nm, and a matched transimpedance limiting amplifier. The demonstrator module was an adaption of an u²t photoreceiver. The setup is schematically depicted in Figure 5.3.

Judging the device performance, highest speed, best energy efficiency, or uncooled operation can be required by the application. To achieve the highest ratings in any of these categories, one has to compromise the others. On the other hand, better devices are usually better in all categories. Consequently, the devices were characterized in

FIGURE 5.3 Setup for characterization. The "quasi-back-to-back" configuration should emulate the application within a short optical interconnect.

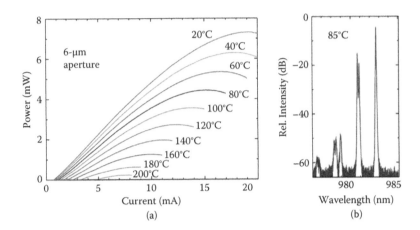

FIGURE 5.4 Static output characteristics: (a) L-I characteristics from 20°C to 200°C. (b) Corresponding spectrum at 85°C. (See References)

top-speed versus temperature [21]. The measurements were performed using butt coupling and a short 3 m multimode fiber link. Typical 6 μm oxide-aperture diameter devices showed threshold currents of 0.9 mA at 20 °C with a differential resistance of 75 Ω and a maximum optical output power exceeding 8 mW at a rollover current of 22 mA. The static characteristics are given in Figure 5.4.

The devices operate continuous wave up to 200 °C ambient temperature [21]. In Figure 5.4a the *L-I* characteristic for a representative VCSEL with a 6 μm oxide aperture at temperatures between 20 °C and 200 °C is depicted. The operation voltage is typically 2 to 3 V. The emission wavelength is between 980 and 985 nm within the temperature range. For 6 μm aperture VCSELs, the spectrum is quasi-single-mode and the shape does not change with temperature, which is important for stable fiber coupling. The spectrum is given in Figure 5.4b. Fiber coupled powers at a constant 6 mA drive current were measured [17] to be 2.3 mW, 1.2 mW, and 200 μW for temperatures of 85 °C, 155 °C and 200 °C, respectively. The threshold current is around 1 mA, slightly increasing with temperature, indicating a too small mode gain offset and room for improvement looking at maximum ambient temperature ratings.

Data-transmission experiments were carried out under ambient temperatures from −14 °C to +155 °C allowing us to identify error-free operation at bit rates from 12.5 to 49 Gb/s. We achieved error-free operation of directly modulated VCSELs with non-return-to-zero (NRZ) coding by a 2^7-1 bits long pseudo-random-bit-sequence (PRBS) at record-high bit rates of 12.5 Gb/s at 155 °C, 17 Gb/s at 145 °C, 25 Gb/s at 120 °C, 38 Gb/s at 85 °C, 40 Gb/s at 75 °C, 44 Gb/s at 25 °C, 47 Gb/s at 0 °C, and 49 Gb/s at −14 °C [1,19,21,22]. Due to the very high temperature stability, longer bit patterns up to PRBS sequences of $2^{31}-1$ showed no or only minor degradation. The results of the large-signal transmission experiments are depicted in Figure 5.5. In Figure 5.5a we show a bit error rate (BER) plot at room temperature demonstrating error-free perfor-mance beyond 40 Gb/s. In Figure 5.5b all large signal experiments are summed up in a temperature versus error-free bit-rate plot and compared with the state of the art

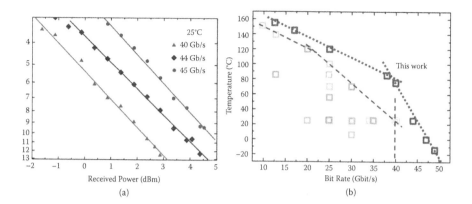

FIGURE 5.5 High-speed modulation of 980-nm VCSELs. (a) Bit-error-rate plots of data transmission experiments at room temperature. (b) Error-free NRZ-bit rate versus ambient temperature including the state-of-the-art in 2010, and the results achieved at the TU Berlin in 2011. (See References)

of 2011. The result presented in Figure 5.5a translates to the 44Gb/s and 25 °C point in Figure 5.5b as we see 44 Gb/s with a BER below 10^{-12} in Figure 5.5a.

This experimental result is in agreement with our theoretical considerations given in Chapter 2. Especially for high-speed VCSELs we believe that a good thermal design is crucial. Consequently, we were able to move the entire border with a single device run. The different slopes toward higher bit rates could be due to limitations (receiver bandwidth below 30 GHz, intended to be used up to 40 Gb/s) on the receiver side.

Even though these lasers are the most energy-efficient devices beyond 30 Gb/s [19], they were clearly optimized for highest bandwidths. Devices optimized for lowest energy consumption per bit are discussed in the next section.

5.4 ENERGY-EFFICIENT 850-NM VCSELS

The standard wavelength for links in the range of some 100 m via multimode fiber is 850 nm. Recently, devices with energy efficiencies below 100 fJ/bit could be demonstrated.

5.4.1 ENERGY-EFFICIENT DATA TRANSMISSION

According to the International Technology Roadmap for Semiconductors (ITRS), lasers for future optical interconnects should be highly energy efficient. In 2015, energy-efficient high-speed lasers operating at 100 mW/Tbps (100 fJ/bit) will be required [11,23]. These numbers refer to the *dissipated electrical energy* per bit to fit the cooling budget of the data center. This figure of merit can be defined as the heat to bit rate ratio (HBR) (mW/Tbps) [24]

$$HBR = P_{\text{diss}} / BR \qquad (5.13)$$

where P_{diss} is the dissipated heat ($P_{\text{diss}} = P_{\text{el}} - P_{\text{optical}}$) of the laser and BR is the bit rate. "Green photonics" means that the *total energy* consumed per transmitted amount of data is of equal importance [24]. Consequently, the electrical energy to data ratio (EDR) (fJ/bit) can be defined as

$$\text{EDR} = P_{el}/\text{BR} \qquad (5.14)$$

where $P_{el} = V \cdot I$ is the total consumed electrical power with V and I as the laser's operating bias point [7]. Additionally, the modulation power absorbed by the laser should be taken into account [19]. Even though, depending on the used electronics, it might be the case that the power actually consumed in the VCSEL is smaller than the power needed by the driving electronics, we believe that it is the most crucial one. This is due to its multiplier effect of the energy consumed by the light source on the power consumption of the whole system. Please note that efficiency per bit is not the same as the wall-plug efficiency (WPE), which can actually be expressed in terms of HBR and EDR:

$$\text{WPE} = 1 - \text{HBR}/\text{EDR} \qquad (5.15)$$

Furthermore, this means that the most power-efficient lasers in terms of data transmission are not necessarily the lasers with the highest WPE, nor is the driving condition for best HBR or EDR identical with the point of highest WPE.

In order to identify the figures of merit for designing an efficient laser looking at the energy per bit, we have to recap the rate equation analysis in Section 5.2. From Equation (5.11), we find

$$f_R^2 \propto S \propto I - I_{th} \qquad (5.16)$$

with the latter proportionality only being valid for small driving currents. Please note that this is especially not the case for strongly biased ultrahigh-speed VCSELs. Assuming linear behavior we can define a D factor and a modified factor D_{BR} based on the bit rate:

$$f_R \equiv D\sqrt{I - I_{th}} \propto BR \equiv D_{BR}\sqrt{I - I_{th}} \qquad (5.17)$$

To model the electrical power consumed by the VCSEL we use an ideal diode with series resistance and define

$$U \equiv U_{th} + R_d I \qquad (5.18)$$

with U_{th} as the threshold voltage and R_d as the differential series resistance.

We can write the EDR as defined in Equation (5.14) as follows:

$$\text{EDR} = \frac{U \cdot I}{\text{BR}} = \frac{1}{\text{BR}}\left[\left(\left(\frac{BR}{D_{BR}}\right)^2 + I_{th}\right)U_{th} + \left(\left(\frac{BR}{D_{BR}}\right)^2 + I_{th}\right)^2 R_d\right] \qquad (5.19)$$

$$= U_{th} \left(\frac{BR}{D_{BR}^2} + \frac{I_{th}}{BR} \right) + R_d \left(\left(\frac{BR^{3/4}}{D_{BR}} \right)^2 + \left(\frac{I_{th}}{\sqrt{BR}} \right)^2 \right) \qquad (5.20)$$

In order to achieve a small value of EDR and an efficient laser, R_D and U_{th} have to be as small as possible. This means we need to optimize for a VCSEL with very low electrical losses. A large bit rate would be required to compensate high threshold currents. On the other hand, for a given D factor, a higher bit rate translates into an inferior efficiency. This is why we suggest using lasers with very small threshold currents. If we neglect the threshold currents in Equation (5.20) we yield:

$$\text{EDR} \mid_{I_{th}=0} = U_{th} \frac{BR}{D_{BR}^2} + R_d \frac{BR^3}{D_{BR}^4} \qquad (5.21)$$

From Equation (5.21) we can learn, that low EDR values become more and more difficult for higher bit rates. A high D factor, on the other hand, is crucial for efficient lasers. The D factor can be written as

$$D = \frac{1}{2\pi} \sqrt{\frac{v_g}{e} \cdot \frac{\eta_i a}{V_{res}}} \qquad (5.22)$$

with η_i as differential quantum efficiency and V_{res} as the volume of the optical resonator. This makes clear that small-aperture VCSELs are beneficial in many ways for energy-efficient links. First, we achieve threshold at low currents and, second, benefit from a larger D factor.

Efficient high-speed VCSELs have been developed in the wavebands from 850 nm to 1550 nm [25] and can be potentially optimized for highest data-transmission efficiencies. In 2011, the topic of green photonics has become of public and scientific interest. Conferences focusing on that topic and scientific awards acknowledging achievements on this field were initiated [26]. VCSELs operating at 10 Gb/s at 1060 nm with 140 mW/Tbps HBR have been reported [27]. Longer wavelengths use less energy per photon and have therefore an intrinsic advantage. Furthermore, active materials with better gain properties can be used [17]. Nevertheless, 850 nm remains the current standard wavelength for fiber-based links. On the other hand, for the application in short-optical interconnects, proprietary solutions with other wavelengths can address also the market. Vast efforts have been made in boosting both bit rate [1,17–19,28–32] and energy efficiency [19,24,26–28,31,33–36] to meet the requirements of future data centers and supercomputers. Researchers in Taiwan could demonstrate single-mode devices with high wall-plug efficiencies and a remarkable HBR of 109 mW/Tbps [34]. Please note that higher bit rates require a quadratic increase in current densities at a given device technology. This makes energy-efficient devices operating at higher bit rates more challenging. EDR of 500 fJ/bit or more are usually needed to achieve bit rates of 30 Gb/s or more [28]. Consequently, at bit rates as high as 35 Gb/s HBR and EDR values in the order of 200 to 300 fJ/bit are also outstanding results [19,33].

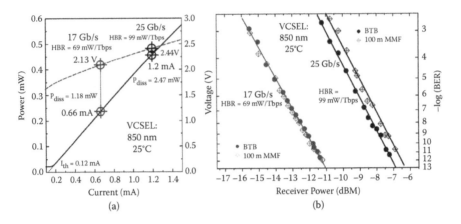

FIGURE 5.6 Energy-efficient 850-nm VCSEL, room temperature. (a) LIV characteristics; the biasing point of the data-transmission experiment is indicated. (b) Bit-error-rate measurement of that device at 17 Gb/s and 25 Gb/s. (Energy efficiency: 17 Gb/s: HBR = 69 mW/Tbps, EDR = 83 fJ/bit; Modulation energy = 10 fJ/bit 25 Gb/s: HBR = 99 mW/Tbps, EDR = 117 fJ/bit; Modulation energy = 6 fJ/bit.) (See References)

5.4.2 ENERGY-EFFICIENT VCSELs FOR INTERCONNECTS

In a first-order approximation, for a given directly modulated VCSEL device, the resonance frequency rises with the square root of the VCSEL power. Therefore it is trivial to understand that high-speed VCSELs typically consume more energy per bit when they are operated at higher bit rates. However, VCSELs designed to work at ultra-high bit rates do not necessarily get more energy efficient just by simply reducing the pumping current and the bit rate. In order to realize energy-efficient high-speed performance, large resonance frequencies must be achieved at a low drive current [35,7].

Recently, achievements in the field of energy-efficient VCSELs were made at the Center of Nanophotonics at the Technical University of Berlin. These devices were optimized for highest energy efficiency. The results for devices in the 850 nm waveband are given in Figure 5.6. The HBR is 69 mW/Tbps at 17 Gb/s and 99 mW/Tbps at 25 Gb/s [24]. A 100-m fiber link shows negligible power penalties. By heating up the device to 55 °C we yield a record-low EDR of 81 fJ/bit and an HBR of 70 mW/Tbps at 17 Gb/s. Data transmission over 1 km of multimode fiber has also been accomplished [35].

Record-high data rates and efficiencies are not achieved at the same driving conditions. On the other hand, as the efficiency of the device is not affected by ambient temperature in a wide range, uncooled systems saving large amounts of energy are feasible.

5.5 ADVANTAGES OF LONG-WAVELENGTH VCSELS FOR INTERCONNECTS

Having seamless solutions and silicon photonics in mind, long-wavelength VCSELs emitting 1.3 μm to 1.6 μm are very attractive light sources. Meanwhile, VCSELs at these wavelengths are also commercially available.

Longer wavelengths need less energy per photon. This translates into lower driving voltages of the active devices. Photon energies below one electron volt make it easier to create energy-efficient CMOS driver chips. Furthermore, silicon becomes transparent at these longer wavelengths. Therefore, this kind of laser source seems to be an ideal candidate for future integrated optics and silicon photonics. Also seamless solutions from a metro-range fiber link via printed circuit boards into the core of a silicon photonics chip could be realized at these wavelengths. On the other hand, the lower energy per photon makes these lasers more vulnerable to device heating. Auger processes become more dominant at higher wavelengths. Free-carrier absorption scales with wavelength at the power of two [16]. Active region and VCSEL chip prefer different base substrates. This makes it more challenging to realize such kind of devices.

5.5.1 CHALLENGES OF LONG-WAVELENGTH VCSELs

With VCSELs, there are two principal issues to conquer that are mostly material related. First, a low-loss, high-q laser cavity has to be realized, and, second, the laser current has to be confined to the active area while concurrently avoiding excessive heating. To reach the lasing threshold, the mirror and cavity losses have to be compensated by the gain of the active laser section. As the overlap of the active region with the optical mode only happens in a thin vertical area, a high-q cavity with high-reflectivity mirrors is needed. GaAs-based devices operating continuous wave at room temperature were already reported in 1988 by Fumio Koyama et al. [37], 11 years after the laser concept had been suggested in 1977 by Kenichi Iga [38]. Realizing VCSELs in the GaAs-material system is relatively easy, as both binary mirrors and stable wet oxidation techniques are available.

Unfortunately, there is no easy solution for realizing long-wavelength VCSELs emitting around the desirable waveband around 1.3 to 1.6 μm:

- Optical losses in p-conducting materials scale with wavelength squared
- On InP heat extraction through quaternary mirror stacks is not efficient
- On GaAs there is no classic material for the active region
- No oxidation of the current aperture on InP
- Very thick epitaxial layers have to be grown precisely and homogeneous

Nevertheless, there has been vast effort in realizing such kind of devices in different concepts. Devices in the InGaAlAs material system, grown on InP substrate have been presented in 1999 [39] (pulsed) and 2000 in CW operation [40,41]. Another approach in realizing long-wavelength VCSEL devices is growing an active region of GaInNAs material (diluted nitrides) on GaAs substrate [42]. This allows using the mature GaAs-based VCSEL technology with its mirrors with good thermal conductivity and oxidized apertures. However, nitrogen-containing materials are not completely understood yet and tend to decompose under certain conditions like extreme temperatures or current densities. VCSELs, on the other hand, are typically driven at rather high current densities and the active region suffers from self-heating. Especially targeting at high modulation speeds, high carrier and photon densities are needed to boost the relaxation oscillation frequency, that is, the intrinsic bandwidth of the laser. Therefore, top performance and reliability are somehow contradictory in this approach and will depend on the quality of the nitrogen containing layers. Active regions based on quantum dots could be the better alternative. Here, the main challenge is the growth of these novel active materials [43].

Another approach is growing the distributed Bragg reflectors (DBRs) on GaAs wafers and the active region on InP. The final layer structure is generated by two wafer-bonding steps [44]. The bonded interfaces, however, show quite bad electrical properties making lateral intracavity contacts necessary [45]. This necessity makes device processing more sophisticated and leads to higher electrical parasitic limiting modulation speeds.

Buried-tunnel-junction long-wavelength VCSELs, which were reported by Ortsiefer et al. in 2000 [40] enabled room temperature CW operation with superior performance.

To sum up, quite different design concepts based on GaAs or InP substrate have matured and are commercially available. Schematics of the competing design concepts are depicted in Figure 5.7.

VCSELs might be the answer to the uprising question of how to quench the never-ending thirst for more bandwidth at lower cost and energy consumption. Due to the challenges as discussed before, several quite different designs of long-wavelength VCSELs have been developed and matured to commercial availability. Industry, however, usually prefers to buy standard products with exceptional performance from a variety of suppliers rather than choosing from proprietary solutions with its pros and cons. Further, recent progress in VCSEL research might have been underestimated. On the other hand, novel applications like silicon photonics requiring these wavebands might change minds.

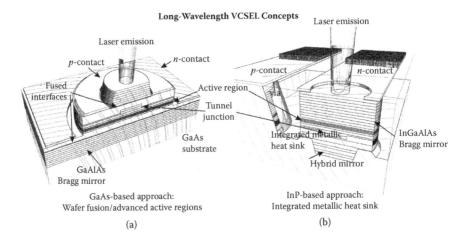

FIGURE 5.7 Long-wavelength VCSELs: (a) GaAs-based devices with active regions based on diluted nitrides or quantum dots or InP-based active regions and wafer-fused GaAs-based mirrors; (b) monolithic concept on InP with a hybrid mirror and integrated metallic heat sink.

REFERENCES

1. W. Hofmann, P. Moser, P. Wolf, A. Mutig, M. Kroh, D. Bimberg, "44 Gb/s VCSEL for optical interconnects," *OFC/NFOEC*, PDPC5, pp. 1–3, 2011.
2. T. Mudge, "Power: A first-class architectural design constraint," *Computer*, 34, pp. 52–58, 2001.
3. F. Doany, C. Schow, C. Baks, D. Kuchta, P. Pepeljugoski, L. Schares, R. Budd, F. Libsch, R. Dangel, F. Horst, B. Offrein, J. Kash, "160 Gb/s bidirectional polymer-waveguide board-level optical interconnects using CMOS-based transceivers," *IEEE Trans. Adv. Packaging*, 32, pp. 345–359, 2009.
4. D. Wiedenmann, M. Grabherr, R. Jäger, R. King, "High-volume production of single-mode VCSELs," *Proc. SPIE*, vol. 6132, pp. 1–12, 2006.
5. A. Vahdat, H. Liu, X. Zhao, C. Johnson, "The emerging optical data center," *OFC/NFOEC*, OTuH2, pp. 1–3, 2011.
6. L. Schares, J. Kash, F. Doany, C. Schow, C. Schuster, D. Kuchta, P. Pepeljugoski, et al., "Terabus: Terabit/second-class card-level optical interconnect technologies," *IEEE J. Sel. Top. Quantum Electron.*, vol. 12, pp. 1032–1044, 2006.
7. W. Hofmann, P. Moser, D. Bimberg, "Energy-efficient interconnects," in *Breakthroughs in Photonics 2011, IEEE Photonics Journal*, 2012.
8. W. Hofmann, "Evolution of high-speed long-wavelength vertical-cavity surface-emitting lasers," *Semicond. Sci. Technol.*, vol. 26, pp. 014011, 2011.
9. S. Chuang, D. Bimberg, "Metal-cavity nanolasers," in *Breakthroughs in Photonics 2010, IEEE Photonics Journal*, p. 288, 2011.
10. E. Wong, M. Mueller, P. Dias, C. Chan, M.-C. Amann, "Energy-saving strategies for VCSEL ONUs," *OFC/NFOEC*, OTu1H5, pp. 1–3, 2012.
11. D. Miller, "Device requirements for optical interconnects to silicon chips," *Proc. IEEE*, vol. 97, pp. 1166–1185, 2009.
12. M. Taubenblatt, "Optical interconnects for high-performance computing," *J. Lightwave Technol.*, vol. 30, pp. 448–457, 2012.
13. L. Coldren, S. Corzine, "Dynamic effects," in *Diode Lasers and Photonic Integrated Circuits*, pp. 184–212, Wiley, New York, 1995.
14. W. Hofmann, "Laser dynamics," in *InP-based Long-Wavelength VCSELs and VCSEL Arrays for High-Speed Optical Communication,* pp. 25–40, Verein zur Förderung des Walter Schottky Institut der Technischen Universität München, Munich, 2009.
15. L. Graham, H. Chen, D. Gazula, T. Gray, J. Guenter, B. Hawkins, R. Johnson, C. Kocot, A. MacInnes, G. Landry, J. Tatum, "The next generation of high-speed VCSELs at Finisar," *Proc. SPIE*, vol. 8276, p. 827602, 2012.
16. J. Buus, M.-C. Amann, *Tunable Laser Diodes and Related Optical Sources*, Wiley-VCH, Weinheim, Germany, 2005.
17. W. Hofmann, P. Moser, P. Wolf, G. Larisch, W. Unrau, D. Bimberg, "980-nm VCSELs for optical interconnects at bandwidths beyond 40 Gb/s," *Proc. SPIE*, vol. 8276, 827605, 2012.
18. Y. Chang, C. Wang, L. Coldren, "High-efficiency, high-speed VCSELs with 35 Gb/s error-free operation," *Electron. Lett.*, vol. 43, pp. 1022–1023, 2007.
19. P. Moser, P. Wolf, A. Mutig, G. Larisch, W. Unrau, W. Hofmann, D. Bimberg, "85 °C error-free operation at 38 Gb/s of oxide-confined 980-nm vertical-cavity surface-emitting lasers," *Appl. Phys. Lett.*, vol. 100, p. 081103, 2012.
20. A. Mutig, *High-Speed VCSELs for Optical Interconnects,* Springer, 2011.
21. P. Wolf, P. Moser, G. Larisch, M. Kroh, A. Mutig, W. Unrau, W. Hofmann, D. Bimberg, "High-performance 980 nm VCSELs for 12.5 Gbit/s data transmission at 155 °C and 49 Gbit/s at −14 °C," *Electron. Lett.*, vol. 48, pp. 389–390, 2012.

22. W. Hofmann, P. Moser, A. Mutig, P. Wolf, W. Unrau, D. Bimberg, "980-nm VCSELs for Optical Interconnects at 25 Gb/s up to 120 °C and 12.5 Gb/s up to 155 °C," *Proc. CLEO/ QELS*, pp. 1–2, 2011.
23. "International technology roadmap for semiconductors," 2007 Edition, http://www.itrs. net/Links/2007ITRS/ExecSum2007.pdf, accessed January 2012.
24. P. Moser, W. Hofmann, P. Wolf, J. Lott, G. Larisch, A. Payusov, N. Ledentsov, D. Bimberg, "81 fJ/bit energy-to-data ratio of 850 nm vertical-cavity surface-emitting lasers for optical interconnects," *Appl. Phys. Lett.,* vol. 98, p. 231106, 2011.
25. A. Larsson, "Advances in VCSELs for communication and sensing," *IEEE J. Sel. Top. Quantum Electron.*, vol. 17, pp. 1552–1567, 2011.
26. P. Moser, J. Lott, P. Wolf, G. Larisch, A. Payusova, G. Fiol, N. Ledentsov, W. Hofmann, D. Bimberg, "Energy-efficient vertical-cavity surface-emitting lasers (VCSELs) for "green" data and computer communication," *Proc. SPIE*, Photonics West, Green Photonics Award in Communications, San Francisco, CA, 2012.
27. S. Imai, K. Takaki, S. Kamiya, H. Shimizu, J. Yoshida, J. Kawakita, T. Takagi, et al., "Recorded low-power dissipation in highly reliable 1060-nm VCSELs for 'green' optical interconnection," *IEEE J. Sel. Top. Quantum Electron.*, vol. 17, pp. 1614–1620, 2011.
28. P. Westbergh, J. Gustavsson, A. Haglund, A. Larsson, F. Hopfer, G. Fiol, D. Bimberg, A. Joel, "32 Gbit/s multimode fibre transmission using high-speed, low-current density 850 nm VCSEL," *Electron. Lett.*, vol. 45, pp. 366–368, 2009.
29. P. Westbergh, J. Gustavsson, B. Kögel, A. Haglund, A. Larsson, A. Mutig, A. Nadtochiy, D. Bimberg, and A. Joel, "40 Gbit/s error-free operation of oxide-confined 850 nm VCSEL," *Electron. Lett.*, vol. 46, pp. 1014–1016, 2010.
30. S. Blokhin, J. Lott, A. Mutig, G. Fiol, N. Ledentsov, M. Maximov, A. Nadtochiy, V. Shchukin, and D. Bimberg, "Oxide confined 850 nm VCSELs operating at bit rates up to 40 Gbit/s," *Electron. Lett.*, vol. 45, pp. 501–503, 2009.
31. T. Anan, N. Suzuki, K. Yashiki, K. Fukatsu, H. Hatakeyama, T. Akagawa, K. Tokutome, M. Tsuji, "High-speed 1.1 μm-range InGaAs VCSELs," *OFC/NFOEC 2008*, OThS5, pp. 1–3, 2008.
32. W. Hofmann, M. Müller, P. Wolf, A. Mutig, T. Gründl, G. Böhm, D. Bimberg, M.-C. Amann, "40 Gbit/s modulation of 1550 nm VCSEL," *Electron. Lett.*, vol. 47, pp. 270–271, 2011.
33. Y. Chang, L. Coldren, "Efficient, high-data-rate, tapered oxide-aperture vertical-cavity surface-emitting lasers," *IEEE J. Sel. Top. Quantum Electron.*, vol. 15, pp. 704–715, 2009.
34. J. Shi, W. Weng, F. Kuo, J. Chyi, "Oxide-relief vertical-cavity surface-emitting lasers with extremely high data-rate/power-dissipation ratios," *OFC/NFOEC*, OThG2, pp. 1–3, 2011.
35. P. Moser, J. Lott, P. Wolf, G. Larisch, A. Payusov, N. Ledentsov, W. Hofmann, D. Bimberg, "99 fJ/(bit·km) energy to data-distance ratio at 17 Gb/s across 1 km of multimode optical fiber with 850-nm single-mode VCSELs," *IEEE Photon. Technol. Lett.*, vol. 24, pp. 19–21, 2012.
36. M.-C. Amann, M. Müller, and E. Wong, "Energy-efficient high-speed short-cavity VCSELs," *Proc. OFC/NFOEC 2012*, OTh4F.1, 2012.
37. F. Koyama, S. Kinoshita, K. Iga, "Room-temperature continuous wave lasing characteristics of GaAs vertical-cavity surface-emitting laser," *Appl. Phys. Lett.*, 55, pp. 221–222, 1989.
38. K. Iga, "Surface-emitting laser—Its birth and generation of new optoelectronics field," *IEEE J. Sel. Top. Quantum Electron.*, vol. 6, pp. 1201–1215, 2000.
39. C. Kazmierski, J. Debray, R. Madani, J. Sagnes, A. Ougazzaden, N. Bouadma, J. Etrillard, F. Alexandre, M. Quillec, "+55 °C pulse lasing at 1.56 μm of all-monolithic InGaAlAs-InP vertical-cavity lasers" *Electron. Lett.*, vol. 35, pp. 811–812, 1999.

40. M. Ortsiefer, R. Shau, G. Böhm, F. Köhler, M.-C. Amann, "Room-temperature operation of index-guided 1.55 μm InP-based vertical-cavity surface-emitting laser," *Electron. Lett.*, vol. 36, pp. 437–438, 2000.

41. W. Yuen, G. Li, R. Nabiev, J. Boucart, P. Kner, R. Stone, D. Zhang, et al., "High-Performance 1.6 μm single-epitaxy top-emitting VCSEL," *Electron. Lett.*, vol. 36, pp. 1121–1123, 2000.

42. H. Riechert, A. Ramakrishnan, G. Steinle, "Development of InGaAsN-based 1.3 μm VCSELs," *Semicond. Sci. Technol.*, vol. 17, pp. 892–897, 2002.

43. N. Ledentsov, F. Hopfer, D. Bimberg, "High-speed quantum-dot vertical-cavity surface-emitting lasers," *Proc. IEEE*, vol. 95, pp. 1741–1756, 2007.

44. A. Syrbu, A. Mircea, A. Mereuta, A. Caliman, C. Berseth, G. Suruceanu, V. Iakovlev, M. Achtenhagen, A. Rudra, E. Kapon, "1.5 mW single-mode operation of wafer-fused 1550 nm VCSELs," *IEEE Photon. Technol. Lett.*, vol. 16, pp. 1230–1232, 2004.

45. A. Mereuta, G. Suruceanu, A. Caliman, V. Iakovlev, A. Sirbu, E. Kapon, "10-Gb/s and 10-km error-free transmission up to 100 °C with 1.3-μm wavelength wafer-fused VCSELs," *Opt. Express*, vol. 17, pp. 12981–12986, 2009.

6 High-Speed Photodiodes and Laser Power Converters for the Applications of Green Optical Interconnect

Jin-Wei Shi

CONTENTS

6.1   INTRODUCTION

Global network data traffic continues to grow, driven primarily by mobile data and Internet video. By taking a look at the network equipment energy consumed breakdown, data center network equipment is expected to become the major power consumer [1]. Recently, the development of optical interconnect (OI) techniques [1–5], which could allow the replacement of the bulky and power-hungry active/passive microwave components with more energy-saving and high-speed optoelectronic devices, has become an attractive choice to further reduce the carbon footprint of data centers and realize the idea of the "Green-Internet" [6,7]. In future supercomputers, the estimated total power consumption and bandwidth for an OI system could be as high as 8 MW and 400 PB/s, respectively [5]. This trend thus greatly drives the development of high-speed light sources and receivers with a further reduction in power consumption.

To see the energy consumption breakdown of a modern OI system [8], the consumpted power in the receiver end, which includes photodiodes, transimpedance amplifiers (TIA), and limiting amplifier (LA), has become the major part of the total consumed power due to the significant achievement in high-speed 850 nm or ~1000 nm vertical-cavity surface-emitting lasers (VCSELs) with ultralow power consumption [9–14] performance in the transmitter side.

In the receiving end of an OI system, the direct current (DC) component of the high-speed optical data signal is usually wasted due to the fact that in traditional p-i-n photodiodes (PDs), reverse bias operation is necessary for high-speed performance, which will result in extra power consumption with excess heat generation. To have the substrate with excellent thermal conductivity in both transmitting and receiving sides of an OI system is thus an important issue for good heat sinking of an OI system.

There are two major trends in the development of photodiodes for the applications of OI system. One is the Si/Ge-based PDs, which can be monolithically integrated with Si-based complementary metal-oxide semiconductor (CMOS) integrated circuits (ICs) and the further reduction in both system size and cost of package can be expected. By use of the Ge-based photoabsorption layer in the waveguide PD structure, excellent high-speed performance for 40 Gbit/sec operation with reasonable responsivity (including waveguide coupling loss) under zero-bias operation has been successfully demonstrated [15,16]. The capability of high-speed and zero-bias operations of the CMOS compatible Ge-based PD is mainly due to that the photogenerated hole in the Ge layer has a much faster drift velocity than that of the hole in the III-V semiconductor-based photoabsorption layer. Such excellent performance indicates its strong potential for green OI application. Furthermore, the Si substrate offers a significantly higher thermal conductivity than that of the GaAs or InP substrates, although the silicon-on-insulator (SOI) substrate, which is usually adopted in Si-photonic technology [17,18], has a poor thermal characteristic due to that the heat flow is blocked by the buried oxide layer. This problem can be minimized by using some poly-Si-based thermal shunt in the layout of chips to directly sink the generated heat from active devices to Si substrates [19]. Figure 6.1 shows a picture of a 4 × 10 Gb/s, 0.13 μm CMOS SOI integrated optoelectronic transceiver chip copackaged with a single, externally modulated CW laser [20].

As can be seen in the figures, the Si-based waveguide, modulator, coupler, and electronic amplifiers are all monolithically integrated on a single chip.

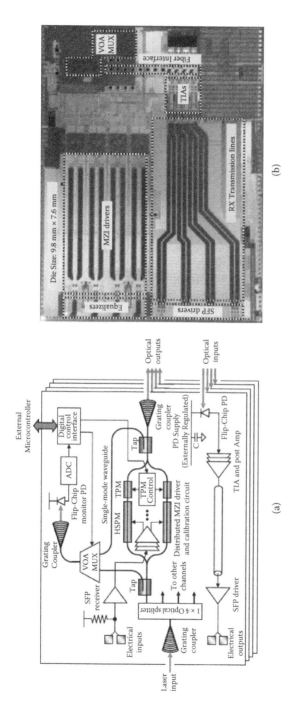

FIGURE 6.1 (a) Conceptual diagram of a 4×10 Gb/s, 0.13 μm CMOS silicon-on-insulator integrated optoelectronic transceiver chip copackaged with a single, externally modulated CW laser; (b) photograph of the same. (From A. Narasimha et al., "A 40-Gb/s QSFP optoelectronic transceiver in a 0.13 μm CMOS silicon-on-insulator technology," *Optical Fiber Communication Conference*, San Diego, CA, February 2008. © 2008 IEEE.)

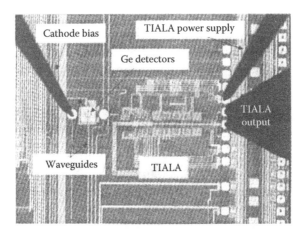

FIGURE 6.2 Photograph of a Ge-based waveguide PD monolithically integrated with a CMOS IC. (From G. Masini et al., "A four-channel 10 Gbps monolithic optical receiver in 130 nm CMOS with integrated Ge waveguide photodetectors," *National Fiber Optic Engineers Conference*, Anaheim, CA, March 2007. © 2007 IEEE.)

These integrated active/passive components (i.e., Si- or Ge-based high-speed photodiodes (PDs) [21–25] that operate at optical wavelengths of 850 nm and 1550 nm) comprise one of the most important active components needed in Si photonic technology. Figure 6.2 shows a picture of a Ge-based waveguide photodiode, which is monolithically integrated with CMOS ICs to form a single chip, for the application to OI or in short-reach fiber communication [24].

The other trend in the development of high-speed PD is to sustain its high-speed performance even when its operating voltage is further pushed to the forward bias. During such operation mode, the PD is functioned as a laser power converter (LPC) [26] and we can thus generate (instead of consuming) DC electrical power by using this device during high-speed optical data detection in an OI system. Figure 6.3 shows the conceptual diagram of the high-speed laser power converter used in the OI system [27]. As can be seen, by use of the bias tee circuits to separate the DC and AC (alternating current) components of detected signal from laser power converter, we can simultaneously generate the DC electrical power in an OI system and obtain clear eye-pattern.

In this chapter, we will review several kinds of high-speed SiGe-based high-speed PDs and III-V-semiconductor-based LPCs for the application of OI. We would introduce their working principles and discuss the factors that limit their bandwidth. Finally, we will also introduce their applications in next-generation OI systems.

6.2 SILICON-BASED HIGH-SPEED PHOTODIODES: 850-NM WAVELENGTH

6.2.1 FUNDAMENTAL PROBLEMS WITH SILICON-BASED HIGH-SPEED PHOTODIODES

Due to the maturity of high-speed 850 nm VCSELs [11,12,14], the application of 850 nm high-speed PDs in OI systems has attracted a lot of attention. Figure 6.4 shows

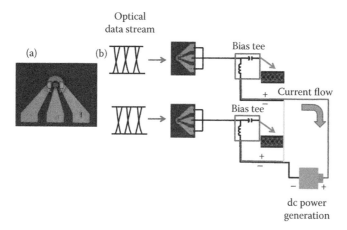

FIGURE 6.3 (a) Top view of a single device and (b) conceptual view of LPC for the OI application. (From J.-W. Shi et al., *IEEE Trans. Electron Devices,* vol. 58, pp. 2049–2056, 2011. © 2011 IEEE.)

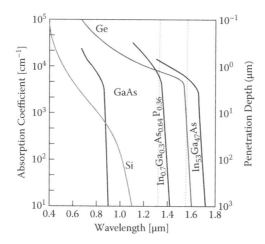

FIGURE 6.4 Photoabsorption coefficients versus wavelengths of different semiconductor materials. (From M. Morse et al., "State of the art Si-based receiver solutions for short reach applications," *Optical Fiber Communication Conference,* San Diego, CA, March 2009. © 2009 IEEE.)

the absorption coefficients and penetration depths versus wavelength for several kinds of semiconductor materials [25], which include Si and Ge. As can be seen, Si has a much weaker absorption coefficient than that of GaAs under 0.8 μm wavelength excitation (1 μm^{-1} versus 0.1 μm^{-1}), which corresponds to a 10 times larger penetration depth into Si material (10 μm versus 1 μm). A larger penetration depth indicates a longer carrier drift time and lower speed performance for Si-based PDs. Furthermore, the depth of the p-n junction is usually on the order of ~1 μm,

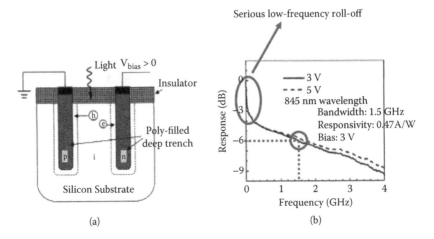

FIGURE 6.5 (a) Conceptual cross-sectional view of a deep-trench p-i-n photodiode; (b) its measured O-E frequency response under 850 nm wavelength excitation. (From M. Yang et al., *IEEE Electron Device Lett.*, vol. 23, pp. 395–397, 2002. © 2002 IEEE.)

which means that most of the photogenerated carriers will be concentrated in the neutral Si substrate without an electrical field inside. Poor speed performance can thus be expected. There are several reported methods to overcome this fundamental problem with Si-based PDs. Figure 6.5a shows a cross-sectional view of a Si-based PD with a deep trench and lateral p-n junction structure [28]. The p-n junction depth (trench depth) is around 7 μm, which is much larger than that of a Si diode with a vertical p-n junction. Figure 6.5b shows its measured optical-to-electrical (O-E) frequency response under the excitations of 845 nm optical wavelengths [28]. As can be seen, under 845 nm wavelength excitation a serious low-frequency roll-off exists and a poor 3 dB bandwidth (<100 MHz, 1.5 GHz 6 dB bandwidth) has been measured. We may thus conclude that even with a deep-trench p-n junction structure (7 μm junction depth) under ~850 nm optical wavelength excitation, the slow diffusion current from the Si substrate seriously limits the 3 dB O-E bandwidth. A larger junction depth (>7 μm) may minimize this phenomenon, however, it will result in a large junction capacitance and poor resistor–capacitor (RC)-limited bandwidth. In the following sections, we will discuss some methods to further overcome such problems.

6.2.2 Spatial-Modulation Silicon-Based Photodiodes at 850-nm Wavelength

It has been demonstrated that the spatial modulation of Si-based PDs, which are fully compatible with the modern CMOS process, diminishes the low-frequency roll-off of Si-based PDs under 850 nm wavelength excitation [29]. Figure 6.6 shows a conceptual cross-sectional view and block diagram of the implementation process. As can be seen, the shaded PD also suffers from a slow diffusion current from the illuminated PD, which results in a low tail in transient response of the PD. However, using a differential amplifier, we can get differential signals for two such PDs

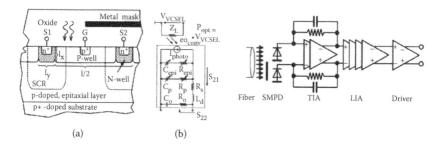

(a)   (b)

FIGURE 6.6 (a) Conceptual cross-sectional view of the spatial modulation of an Si-based p-i-n photodiode; (b) its setup for high-speed operation. (From M. Jutzi et al., *IEEE Photon. Technol. Lett.*, vol. 17, pp. 1268–1270, 2005. © 2005 IEEE.)

(shaded and illuminated), which partially cancel the long tail (low-frequency roll-off) during the transient response. However, the magnitude of the output differential signal is less than that of a single photodiode. The long tail of the illuminated PDs cannot be 100% canceled out due to the unequal distribution of substrate photocurrents from the illuminated and shaded PDs. A 2 Gbit/sec data transmission at an optical wavelength of 850 nm has been demonstrated using such a technique.

6.2.3 SILICON-BASED HIGH-SPEED AVALANCHE PHOTODIODES THAT OPERATE AT THE 850-NM WAVELENGTH

Recently, several research groups have fabricated high-speed Si-based PDs on standard Si substrates by biasing these PDs under avalanche operation [30–33]. The high-speed and high-gain avalanche current screens the slow diffusion current that arises from the Si substrate (as discussed earlier), leading to improvement in the bandwidth. Figure 6.7 shows a conceptual cross-sectional view of Si-based PDs with an interdigitated p-n junction structure and its bias-dependent speed performance [30]. As can be seen, the speed performance of the PDs increases significantly with the reverse bias voltage once the bias voltage reaches the breakdown point. Furthermore, under avalanche operation, a resonant phenomenon may appear in the measured O-E frequency response, which induces a further increase in the measured O-E bandwidth [31–33]. This phenomenon has been attributed to the space-charge effect [33] or the traveling space-charge wave [31]; ultrahigh gain-bandwidth products have also been reported [31,33]. Figure 6.8 shows the measured O-E frequency response reported for Si/SiGe-based APDs under avalanche operation and 850 nm optical wavelength excitation [31].

A significant resonant phenomenon can be observed. By use of this phenomenon, clear 10 Gbit/sec eye opening with reasonable sensitivity performance (–11 dBm) for error-free operation can be achieved under 850 nm optical wavelength excitation [31]. Recently, another research group utilized the structure of Si-based spatial-modulation PDs [34], as discussed earlier, operated in the avalanche region, to realize monolithic Si-based photoreceiver circuits that operate at the 850 nm optical wavelength. These circuits have improved responsivity and speed performance. 10 Gbit/sec eye-opening and error-free operation can also be achieved [34].

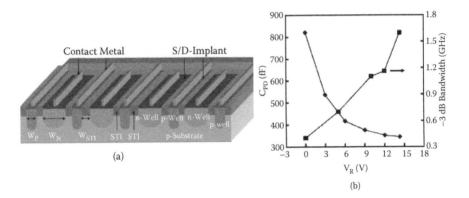

(a)

(b)

FIGURE 6.7 (a) Conceptual cross-sectional view and top view of interdigitated CMOS-based PDs; (b) measured 3 dB bandwidth and junction capacitance versus reverse bias voltage. (From W.-K. Huang et al., *IEEE Photon. Technol. Lett.*, vol. 19, pp. 197–199, 2007. © 2007 IEEE.)

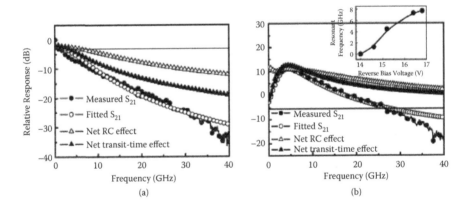

FIGURE 6.8 Measured O-E frequency responses of Si–SiGe APD (a) below and (b) above breakdown voltage operation. (From J.-W. Shi et al., *IEEE Electron Device Lett.*, vol. 30, pp. 1164–1166, 2009. © 2009 IEEE.)

6.3 GERMANIUM AND GERMANIUM-SILICON BASED HIGH-SPEED PHOTODIODES FOR 1.3- TO 1.55-MM WAVELENGTH OPERATION

6.3.1 FUNDAMENTAL PROBLEMS OF GERMANIUM AND GERMANIUM-SILICON–BASED HIGH-SPEED PHOTODIODES

The 1.3 and 1.55 μm wavelengths attract more attention than the 850 nm optical wavelengths due to the fact that in those wavelength regimes the propagation loss and dispersion in the optical fiber is lower. As shown in Figure 6.4, the absorption of the bulk Si material is negligible in that wavelength (1.3–1.55 μm) regime. Using Ge is a promising way to extend the capability of long-wavelength detection of Si-based high-speed PDs. As can be seen in Figure 6.4, the photoabsorption constant of the bulk Ge material is comparable to that of the III-V-based semiconductor material (InGaAsP), under 1.3 μm wavelength operation. However, when the operating wavelength reaches 1.55 μm, the absorption constant of Ge dramatically decreases and becomes much smaller than that of the III-V-based material (InGaAs). The absorption edge of the bulk Ge material seems to be around 1.5 μm. In addition, there is a large difference between the lattice constants of the Ge and Si materials, which results in problems with the growth of high-quality Ge thin films on Si substrates. In the following sections, we will describe several reported technologies to overcome the aforementioned problems.

6.3.2 HIGH-SPEED GERMANIUM-BASED PHOTODIODES

Ge substrates can be used to fabricate high-speed Ge-based metal-semiconductor-metal (MSM) or p-i-n PDs. Figure 6.9 shows the top view, cross-sectional view, and measured external quantum efficiency versus wavelength of Ge PDs [35]. One of the major problems with Ge-based PDs compared with InP-based PDs is their higher dark current density. The proper choice of contact metal, such as silver as is shown in Figure 6.9, is one possible solution to enhance the barrier height and reduce the dark current density. In addition, dopant segregation in MSM Ge-based PDs has been demonstrated to lead to a higher Schottky barrier height and lower dark current [36]. Aside from the high dark current, another problem with Ge-based PDs is the cutoff of the photoabsorption under 1.55 μm wavelength excitation. As can be seen in Figure 6.8, serious external efficiency roll-off is measured when the wavelength reaches 1.55 μm. Applying strain to the Ge epilayer is one possible way to further extend the cutoff wavelength [37].

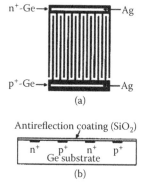
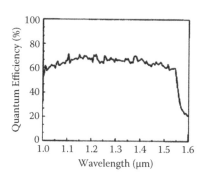

FIGURE 6.9 (a) Top view and (b) cross-sectional view of Ge-based PDs and measured external quantum efficiency versus wavelength. (From J. Oh et al., *IEEE Photon. Technol. Lett.,* vol. 14, pp. 369–371, 2002. © 2002 IEEE.)

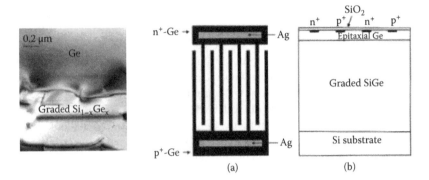

FIGURE 6.10 Cross-sectional view of Ge-based PDs with a graded SiGe buffer on an Si substrate. (From J. Oh et al., *IEEE J. Quantum Electron.*, vol. 38, pp. 1238–1241, 2002. © 2002 IEEE.)

This kind of PD will be discussed in more detail in the following section. Another important issue with the Ge-based PD is its capability of integration with Si-based ICs. Figure 6.10 shows a cross-sectional view of a Ge-based PD grown on a Si substrate [38]. As can be seen, a thick SiGe buffer (several micrometers) is necessary due to the large lattice mismatch between the Ge and Si layers. This thick buffer layer impedes the monolithic integration of Si-based ICs on the bottom layer and Ge-based PDs on the top. The special buffer-free growth condition of the Ge layer on the Si-substrate is one promising solution to overcome this problem. We will discuss this issue in the following section.

6.3.3 HIGH-SPEED AND HIGH-POWER GERMANIUM-ON-SILICON PHOTODIODES

As shown in Figure 6.10, due to the large lattice mismatch (4%) between the Si and Ge material, a thick SiGe buffer layer is necessary for full Si wafer growth. To significantly lower the growth temperature of a thin Ge seed layer, which is directly grown on Si substrate, followed by a thick, high-temperature Ge layer, is one attractive solution to the realization of a high-quality, buffer-free Ge epilayer on the Si substrate [15–17,22]. During the growth process, the growth temperature is lowered down to around ~300°C to produce a thin (several nm), flat Ge seed layer covering the Si-substrate followed by a thick Ge epilayer (several micrometers in thickness) grown at a high temperature (500°C–700°C) [15–17,22]. After that, cyclical high-low temperature hydrogen gas annealing (900°C–700°C) may be performed to further reduce the threading discolation and grow a high-quality Ge layer on the Si substrate [17]. By use of such a technique, to directly grow the Ge thin film on an 8-inch Si wafer is possible for wafer scale device processing [22].

In order to further reduce the threading dislocation defect density, combing the selective area growth technique [15–17] with the aforementioned growth conditions is one attractive solution. In this solution, the Si wafer is patterned by a thick field oxide layer with photolithography-defined open window for Ge layer growth [15]. During the growth process, the dislocation defects would terminate on the window-edge (sidewall) and the defect density on normal Ge surface can thus be reduced. However, the Si plane for Ge layer selective-growth process is usually on the (111) direction, which would result in a mountain shape of grown Ge layer and an additional chemical-mechanical polish (CMP) planarization process is thus necessary [15].

Another important advantage of the Ge layer directly grown on Si substrate is its special characteristic of strain, which would enhance the photoabsorption process. Ge on Si should be compressively strained when grown coherently because of the larger lattice constant of Ge layer. However, the thick Ge epilayer on Si substrate shows tensile strain in the epilayers of the PDs. This is because the linear lattice expansion coefficient of Ge is larger than that of Si, provided the epilayer is not relaxed during growth cooling [17]. This tensile-strained Ge epilayer can extend the absorption cut-off wavelength of the Ge layer to around 1.55 μm. Figure 6.11 shows the measured absorption coefficient versus wavelengths for the Ge layer, with and without tensile strain [37]. As can be seen, the area of selective growth in the Ge layer will have residual tensile strain (0.141%) and enhanced photoabsorption on the long wavelength side compared to that of the full-relaxed Ge layer. The value of the absorption constant for the strained Ge layer is comparable to that of the III-V-based $In_{0.53}Ga_{0.47}As$ layer. Very high-speed performance (>50 GHz) by the selective area Ge PD growth on a Si substrate has been demonstrated at a 1.55 μm wavelength [15,16,22].

Another advantage of the Ge-on-Si growth technique is that it makes possible the realization of high-quality GeSi heterostructures. One of the most important motivations for the realization of GeSi heterostructures is for application in high-performance avalanche photodiodes (APDs) at around the 1.55 μm wavelength regime. This is because the Si material has the largest difference in ionization coefficients between electrons and holes among all the III-V-based semiconductor materials. However, the bandgap of Si restricts its application as high-performance APDs in the fiber communication wavelength regime. By combining the buffer-free grown strained Ge photoabsorption layer, which has reasonable photoabsorption at the 1.55 μm wavelength, and the Si-based multiplication layer, a high-performance telecommunication APD can thus be expected. Figure 6.12 shows the conceptual cross-sectional view of a GeSi APD [39,40]. As can be seen, the high-quality Ge absorption layer is directly

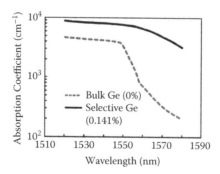

FIGURE 6.11   The absorption constants of tensile-stained Ge film and full-relaxed Ge film on Si substrate. (From H.-Y. Yu et al., *IEEE Electron Device Lett.*, vol. 30, pp. 1161–1163, 2009. © 2009 IEEE.)

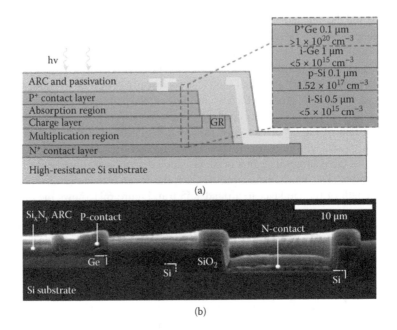

(a)

(b)

FIGURE 6.12 (a) Conceptual cross-sectional view; (b) SEM image of the Ge-on-Si APD. (From M. Piels et al., "Microwave nonlinearities in Ge/Si avalanche photodiodes having a gain bandwidth product of 300 GHz," *Conference on Optical Fiber Communication*, San Diego, CA, March 2009. © 2009 IEEE.)

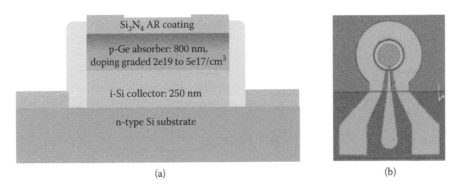

(a)   (b)

FIGURE 6.13 (a) Cross-section schematic and (b) optical photograph of the fabricated device. (From M. Piels and J. E. Bowers, *Opt. Express*, vol. 20, no. 7, pp. 7488–7495, 2012. © 2012 OSA.)

grown on the Si-based multiplication layer. This kind of APD has demonstrated a very high gain-bandwidth product at the 1.3 μm wavelength [39,40].

In addition, another attractive point of Ge-on-Si PDs is its high-saturation output power performance due to the superior thermal conductivity of n-type Si substrate as compared to those of other III-V-semiconductor substrates. Figure 6.13 shows the conceptual cross-sectional view and top view of the proposed GeSi unicarrier traveling carrier photodiode (UTC-PD) [41]. By use of the p-type Ge-based photoabsorption layer, the photogenerated hole would relax to the p-type ohmic contact directly without transport. On the other hand, the photogenerated electron would diffuse through the p-type absorption layer and drift cross the i-Si collector layer, which is transparent under 1.55 μm optical wavelength excitation. Due to that only electron being the active carrier in such a structure and good thermal conductivity of the Si substrate, an excellent high-power performance at ~1.55 μm optical wavelength regime has thus been demonstrated [41]. Figure 6.14 shows the measured bias dependent O-E frequency responses and measured 3 dB O-E bandwidths under different output photocurrents [41]. As can be seen, such Si-based UTC-PD can sustain its high-speed performance (~20 GHz 3 dB bandwidth) under a high output photocurrent (~15 mA) and moderate reverse bias voltage (–3 V). Figure 6.15a shows the measured photogenerated RF power at 20 GHz versus photocurrent under different reverse bias voltage. Figure 6.15b shows the compression

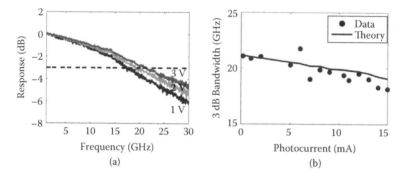

FIGURE 6.14 (a) Bandwidth as a function of bias voltage at 200 μA photocurrent. (b) Bandwidth at 3 V as a function of photocurrent. (From M. Piels and J. E. Bowers, *Opt. Express*, vol. 20, no. 7, pp. 7488–7495, 2012. © 2012 OSA.)

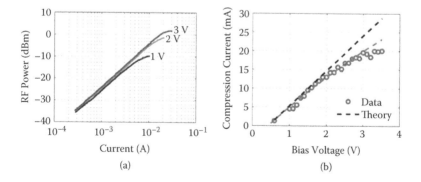

FIGURE 6.15 (a) Output RF power at 20 GHz as a function of photocurrent. (b) 1 dB compression current as a function of bias voltage. Circles: measured data; upper line: 1 dB compression current predicted by model without thermal effects; lower line: model with thermal effects. (From M. Piels and J. E. Bowers, *Opt. Express*, vol. 20, no. 7, pp. 7488–7495, 2012. © 2012 OSA.)

current versus reverse bias. As can be seen, under –3 V bias, the output saturation (compression) current can be as high as 20 mA. Although the demonstrated high-power performance of SiGe UTC-PD is poorer than those of InP-based UTC-PDs [42] due to the smaller electron drift velocity in the Si collector layer, the demonstrated high-power performance of SiGe UTC-PD is superior to the high-power performance of commercial available InP-based p-i-n PD. This excellent high-power performance implies the applications of SiGe UTC-PD for analog and radio-over-fiber communication systems.

6.3.4 PHOTODIODES BASED ON GERMANIUM-ON-INSULATOR (GOI) AND SILICON-ON-INSULATOR (SOI)

An alternative to the fabrication of high-speed Si-based PDs on the standard Si substrate, is the fabrication of high-speed PDs on silicon-on-insulator (SOI) or Ge-on-insulator (GOI) substrates, both of which show some unique advantages. Figure 6.16 and Figure 6.17 show the conceptual cross-sectional view of high-speed

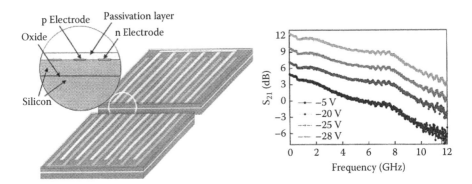

FIGURE 6.16 Conceptual cross-sectional view of an SOI PD and its measured O-E frequency response under 850 nm wavelength excitation. As can be seen, there is no low-frequency roll-off. (From B. Yang, et al., *IEEE Photon. Technol. Lett.*, vol. 15, pp. 745–747, 2003. © 2003 IEEE.)

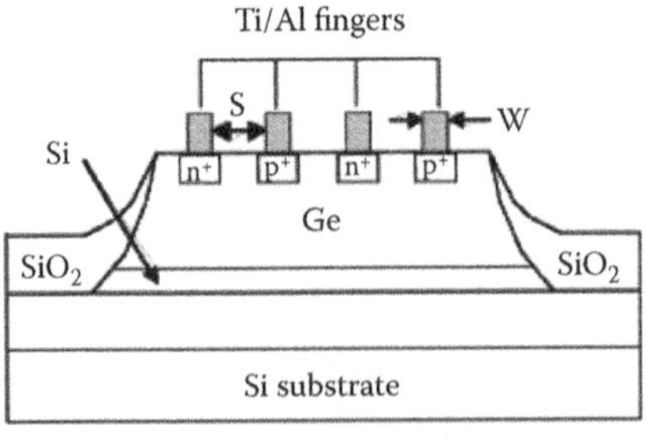

Ge-on-SOI Photodiode

FIGURE 6.17 Conceptual cross-sectional view of the GOI PD. (From C. L. Schow et al., *IEEE Photon. Technol. Lett.*, vol. 18, pp. 1981–1983, 2006. © 2006 IEEE.)

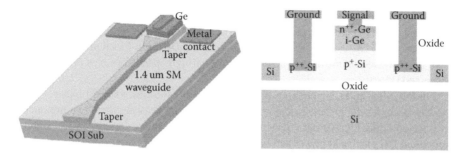

FIGURE 6.18 Conceptual cross-sectional and top views of a GOI based waveguide PD (From T. Yin, *Opt. Express,* vol. 15, no. 21, pp. 13965–13971, 2007. © 2007 OSA.)

PDs on SOI and GOI substrates, respectively [21,23]. The major advantage of SOI and GOI PDs compared to those on the standard Si substrates is that the slow diffusion current from the Si substrate (under 850 nm optical wavelength excitation) is blocked by the buried insulator layer. The shaded PDs structure and avalanche operation of the PD on a standard Si substrate, as discussed in Section 6.2.2 and Section 6.2.3, may not be necessary for minimizing the influence of slow substrate current on the speed performance, in the case of SOI and GOI substrates. Furthermore, SOI and GOI substrates also play important roles in the realization of chip-level optical interconnects and Si-based photonic-integrated circuits (PICs). As shown in Figure 6.1, the integration of active PDs and Si-based modulators with passive optical waveguides on the same Si-chip to form Si-based PICs is necessary. SOI or GOI substrates offer a promising solution for the realization of such systems. This is due to the fact that there is a huge difference in the index between the buried oxide layer and active Ge or Si epilayers. A strong index guided ride waveguide on SOI or GOI substrates can thus be expected. Figure 6.18 shows conceptual top views and cross-sectional views of a GOI waveguide PD for chip-level optical interconnect applications. As can be seen, the topmost Ge layer serves as a photoabsorption layer to absorb photons propagating at the 1.55 μm wavelength in the SOI waveguide. Ultrahigh speed (>40 Gbit/sec) performance with reasonable responsivity (~0.4 A/W, including 3 dB optical coupling loss for the input SOI waveguide) and low dark current (~3 nA) can be achieved at a 1.55 μm wavelength under zero-bias operation, which is very suitable for the application of a green OI system [15,16] due to its zero power consumption. Such excellent zero-bias speed performance of Ge-based p-i-n waveguide PD can be attributed to the fast drift velocity of hole in the Ge photoabsorption layer. This performance is even comparable with the high-performance III-V-based UTC-PDs under zero-bias operation [42].

6.4   III-V-SEMICONDUCTOR-BASED HIGH-SPEED LASER POWER CONVERTER

6.4.1   P-I-N PHOTODIODE OPERATED UNDER FORWARD BIAS

As aforementioned Si-based PD, a high reverse bias voltage (over –3 V) is usually necessary to deplete the deep Si absorption layer (due to its small photoabsorption constant), accelerate the drift process of electron, and achieve high-speed performance. However, a high DC reverse bias and output photocurrent would result in significant power consumption in the OI system. To incorporate the high-quality Ge-based absorption layer with the Si waveguide is one attractive solution for high-speed zero-bias operation, as discussed [15,16].

On the other hand, by pushing the DC bias of PD into the forward bias regime, we can generate the DC electrical power instead of just wasting the DC component of injected optical power onto device. This device would function as an LPC and its generated DC electrical power can be reused in the OI system and improve the overall efficiency of system. However, forward bias would kill the speed performance of traditional p-i-n PD due to that the extremely small electric field under forward bias would slow the drift velocity of the carrier (photogenerated hole) and increase the junction capacitance.

In this section, we will review our recent work on III-V-semiconductor-based high-speed laser power converter for application in a green OI system [27,44]. By use of the GaAs/In$_{0.5}$Ga$_{0.5}$P LPC structure, we can achieve 10 Gbit/sec error-free data transmission and DC electrical power generation with 20% efficiency.

6.4.2 GaAs-Based High-Speed Laser Power Converters

Recently, the InP-based UTC-PD structure has been used to achieve reasonable high-speed and responsivity performance under zero-bias operation [39]. By use of such structure, we can thus eliminate the excess power consumption of PD during high-speed data detection. The detail working principle of UTC-PD was introduced in Section 6.3.3. In contrast to the traditional p-i-n PD, the slow hole transport can be eliminated in UTC-PD structure, and its high-speed performance can be sustained even under forward (zero) bias regime, where the electric filed in the depletion layer is weak (<~10 kV/cm). Although the InP-based UTC-PD has demonstrated excellent speed performance at 1.55 μm wavelength, under 0.8 μm wavelength excitation, which is the most popular wavelength for OI application, the incident photons produce enough photon energy to induce absorption in the whole epilayer structure. Thus in our previous work, we have demonstrated a high-speed GaAs/AlGaAs-based LPC, which is composed of a GaAs based p-type photoabsorption layer and an Al$_{0.15}$Ga$_{0.85}$As-based collector layer to avoid the undesired photoabsorption that occurs in the collector layer under 800 nm wavelength excitation [27].

Figure 6.19 shows a cross-sectional view of the demonstrated LPC. We adopted the typical vertical-illuminated PD structure with an active circular mesa and a p-type ring contact on the top. The diameter of the whole mesa and the inner circle for light illumination were 28 μm and 20 μm, respectively. As shown in Figure 6.19, the epilayer structure of our device is similar to that reported for our GaAs/AlGaAs-based UTC-PD [45] at the 850 nm wavelength. The device mainly consists of a p-type GaAs-based photoabsorption layer with a 450nm thickness and an un-doped Al$_{0.15}$Ga$_{0.85}$As based collector layer with a thickness of 750 nm. A graded p-doped profile (1 × 10^{19} cm^{-3} [top] to 1 × 10^{17} cm^{-3} [bottom]) is used in the absorption layer to accelerate the diffusion velocity of the photogenerated electrons. Compared with the traditionally structured p-i-n PD, we can expect much higher speed performance

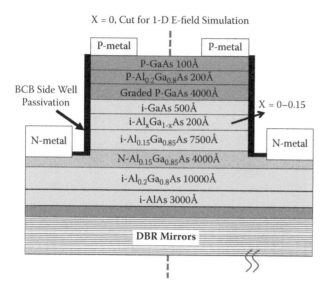

FIGURE 6.19 Cross-sectional view of the demonstrated high-speed GaAs/AlGaAs-based LPC. (From J.-W. Shi et al., *IEEE Trans. Electron Devices,* vol. 58, pp. 2049–2056, 2011. © 2011 IEEE.)

under zero or forward bias operation because there are only electrons in the active carrier. This means the device can exhibit a much faster drift velocity than that of the hole in p-i-n PD under a small electric field (~10 kV/cm) [42]. As can be seen in Figure 6.19, the whole epilayer structure of the LPC was grown on an n-type distributed Bragg reflector (DBR) to enhance its responsivity performance. The main drawback of the photovoltaic LPC is its low output voltage, which is usually too low to directly power other active components in the OI system. In order to boost the DC operating voltage of our LPC, several high-speed LPCs are series wound (linear cascade). The DC operation voltage would thus be proportional to the number of cascaded units. In addition, in order to maximize the net output photocurrent, the output photocurrent from each cascaded unit must be as close as possible [26].

The measured DC responsivity of our single device is around 0.41 A/W, which corresponds to around an external quantum efficiency of 60% under zero bias operation. Under low power injection (<0.5 mW), there is a slight degradation in this value to around 0.36 A/W when the operating voltage reaches +0.9 V. Figure 6.20 shows the measured current (I)–voltage (V) curves of the single LPC and linear cascade two-LPC devices under different output photocurrents. The value of the injected optical power to each single device is specified. As can be seen, the operating voltage of the cascade device is about two times higher than that of a single cell with around one-half of the responsivity. This is because the optical power needs to be about two times higher to feed each single cell in the cascaded structure to generate the same amount of output photocurrent as that of a single device. Figure 6.21 shows the measured O-E power conversion efficiency versus bias voltage of the single and cascaded two-LPC devices under different optical pumping powers. The maximum O-E

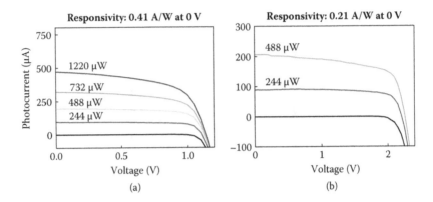

FIGURE 6.20 Measured I–V curves of (a) single device and (b) linear cascade device under different optical pumping powers. (From J.-W. Shi et al., *IEEE Trans. Electron Devices*, vol. 58, pp. 2049–2056, 2011. © 2011 IEEE.)

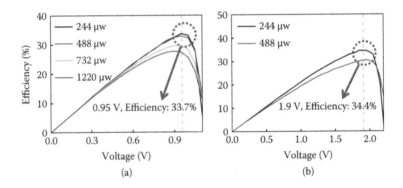

FIGURE 6.21 Measured O-E power conversion efficiency versus bias voltage of the (a) single device and (b) linear cascade device under different optical pumping powers. (From J.-W. Shi et al., *IEEE Trans. Electron Devices*, vol. 58, pp. 2049–2056, 2011. © 2011 IEEE.)

power conversion efficiency of both devices under low pumping power and optimum bias voltage is around 34%. On the other hand, when the injected optical power exceeds 0.5 mW, as shown in Figure 6.20 and Figure 6.21, significant reduction in both the photocurrent and external efficiency can be observed.

Figure 6.22 shows the measured O-E and extracted RC/transient-time limited 3 dB bandwidths of single device and two-element cascaded device under different forward bias. When the forward operation voltage further increases, the 3 dB bandwidth performance seriously degrades due to the significant increase in carrier transient time. This phenomenon can be attributed to the current blocking effect in the GaAs/AlGaAs heterojunction, which can be minimized by use of type-II heterostructures at the interface between the absorption and collector layers [46]. Furthermore, the cascaded device has a superior speed performance compared to that of single device. This result can be attributed to an imbalance in the optical

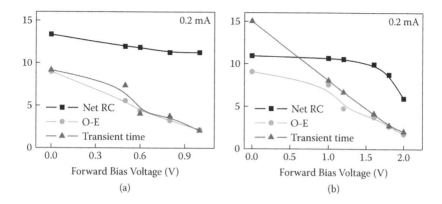

FIGURE 6.22 Measured O-E, extracted RC and transient time limited 3 dB bandwidths versus forward bias voltages for the (a) single device and (b) cascaded device. (From J.-W. Shi et al., *IEEE Trans. Electron Devices,* vol. 58, pp. 2049–2056, 2011. © 2011 IEEE.)

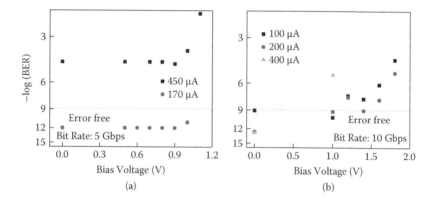

FIGURE 6.23 Measured BER versus forward bias voltages of the (a) single device at 5 Gbit/sec and (b) linear cascade device at 10 Gbit/sec under different output photocurrents. (From J.-W. Shi et al., *IEEE Trans. Electron Devices,* vol. 58, pp. 2049–2056, 2011. © 2011 IEEE.)

power injected into the two-cascaded devices (1:1.2), which causes the device with the lower injected optical power to suffer from a smaller effective DC or instantaneous AC forward voltage and a larger built-in field compared with the device with a higher injected power. The larger built-in field should avoid the slow down of the overshoot electron drift velocity and a higher 3 dB bandwidth.

Figure 6.23 shows the measured bit error rate (BER) versus forward bias voltage under different output photocurrents for the single device at 5 Gbit/sec and linear cascade device at 10 Gbit/sec (pseudo random bit sequence, PRBS: $2^{15}-1$). Figures 6.24 and 6.25 shows the corresponding error-free (BER $<1 \times 10^{-9}$) eye patterns at 5 and 10 Gbit/sec under different forward bias voltages for the single and cascaded devices, respectively. We can clearly see that the maximum operation speed of single device is just around 5 Gbit/sec and by use of the cascade structure,

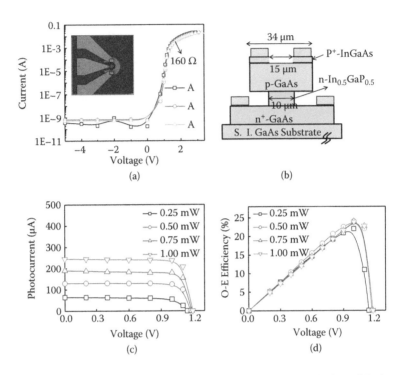

FIGURE 6.24 The (a) dark I–V curves, (b) conceptual cross-sectional view of device with undercut mesa, (c) photocurrents measured under different optical pumping powers and forward voltages for devices A (open symbols), and (d) corresponding DC O-E power conversion efficiency. The inset to (a) shows the top view of the demonstrated LPC.

the operation speed can be further improved to 10 Gbit/sec. The faster speed performance of the cascaded device can be mainly attributed to the shorter internal carrier transient time, as discussed.

Although the cascaded device can attain 10 Gbit/sec operation, the certain ratio of optical injected power onto the cascaded two devices are necessary, which limits its practical application in OI system. The key issue to further improve the speed performance of LPC is to minimize the current blocking effect.

In order to attain such a goal, the GaAs/In$_{0.5}$Ga$_{0.5}$P LPC has been demonstrated [44]. The inset in Figure 6.24a and Figure 6.24b shows a top view and conceptual cross-sectional view of the fabricated device, respectively. The main difference between such a structure and the discussed GaAs/Al$_{0.15}$Ga$_{0.85}$As-based LPC is that the AlGaAs-based collector layer is replaced by a 730 nm In$_{0.5}$Ga$_{0.5}$P n-type collector (C) layer with a graded doping profile (1×10^{16} cm^{-3} [top] to 5×10^{18} cm^{-3} [bottom]). The graded doping profile induced built-in electric (E) field in C layer can accelerate the electron diffusion/drift process, which would significantly benefit the high-speed performance of our LPC operated under forward bias with a very small net E-field inside. Although the use of graded n-type doping in the collector layer would increase the junction capacitance and degrade the RC-limited bandwidth, compared with those of a device with an undoped collector layer [27], such a problem can be minimized by using the $In_{0.5}Ga_{0.5}P$ mesa with an undercut structure, as will be discussed later. Furthermore, the $In_{0.5}Ga_{0.5}P$ collector layer can greatly minimize the conduction band offsets between the GaAs-based absorption and C layers, which would block the electron current and seriously limits the speed performance of LPC [27], as discussed.

To further reduce junction capacitance without seriously reducing the device active area and increasing the differential resistance, an undercut mesa profile (as shown in Figure 6.24b) has been realized in our device [44]. By properly controlling the wet-etching time, the active diameter of the final fabricated device is around 10 μm, as specified in Figure 6.24b. The detail fabrication processes of our device can be referred to in our previous work [44].

Figure 6.24a shows the measured I–V curves of device with undercut mesa profile. As can be seen, the differential resistance is around ~160Ω, which is close to the reference device without undercut profile. The measured DC photocurrent of our LPC versus the forward voltage under different optical pumping power with wavelength at 830 nm are shown in Figure 6.24c and the corresponding O-E power conversion efficiency is shown in Figure 6.24d. As can be seen, the maximum O-E conversion efficiency happens at the bias around +1.0 V and the corresponding maximum conversion efficiency is ~23%. A higher conversion efficiency of our demonstrated LPC can further be expected by inserting a distributed Bragg reflector with central wavelength around ~830 nm below the active layers of device, as shown in Figure 6.23, to enhance its photoabsorption process.

Figure 6.25 shows the measured O-E response (f_{O-E}) of device with undercut (device A) and reference device without undercut (device B) under a fixed output photocurrent (~90 μA) and two different operating voltages (–5 V and +0.8 V). Under +0.8 V operation, the corresponding maximum conversion efficiency is up to ~20%. As can be seen for device A, even when the operating voltage is pushed to the near turn-on point, the degradation in speed is not so serious (10 to 8 GHz). On the other hand, device B exhibits a more serious degradation in speed performance (9 to 2 GHz) due to its larger junction capacitance than those of device A.

The influence of forward operation voltage on the measured high-speed eye patterns is a key issue of demonstrated LPC for practical application. Figure 6.26 shows the

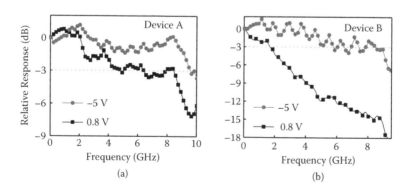

FIGURE 6.25 Measured O-E frequency responses of (a) device A and (b) device B under +0.8 V and –5 V under a fixed photocurrent (90 μA).

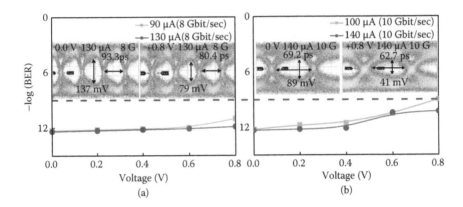

FIGURE 6.26 Measured BER versus forward operation voltages of device A with different output photocurrents at (a) 8 Gbit/sec and (b) 10 Gbit/sec operations. The insets show the corresponding error-free eye patterns.

measured BER of device A versus forward operation voltage under 8 and 10 Gbit/sec operations. The insets show the corresponding error-free (BER $<10^{-9}$) eye patterns. As can be seen, under 8 Gbit/sec operations, the error-free operation can be sustained from 0 to +0.8 V, which corresponds to 20% DC O-E power generation efficiency. As shown in Figure 6.26b, even though the data rate is up to 10 Gbit/sec, error-free (BER $<10^{-9}$) performances can still be maintained when the operation voltage is pushed to +0.8 V.

6.5   CONCLUSION

In this chapter, we have reviewed several kinds of high-speed PDs and III-V-based high-speed LPCs for green OI applications. Under 850 nm wavelength excitation, the main problem is the slow diffusion current from the Si substrate, which will seriously degrade the speed performance of the PD. In order to overcome this problem, a shaded PD structure, avalanche operation, and SOI substrates are utilized to block or screen this slow substrate photocurrent. A Ge-based epilayer can be incorporated into the Si substrate for 1.3 to 1.55 μm wavelength detection. A special growth conditions and cycling temperature annealing process are necessary to overcome the problem caused by the 4% lattice mismatch between the Ge and Si layers. Very high speed performance (40 Gbit/sec) of Ge-on-Si-based PDs and GOI PDs and APDs have been demonstrated at 1.3 and 1.55 μm under zero bias operation. Furthermore, due to the excellent thermal conductivity of Si substrate, the high-power and high-speed performance of GeSi UTC-PD has also been demonstrated.

Except for the significant progress in the development of PDs for the OI application, the development of III-V-semiconductor-based PDs is to further push its operation voltage to forward bias and let it function as a laser power converter. By use of the type-II UTC-PD structure to eliminate the problem of hole transport and current blocking effect, the demonstrated device can achieve invariable high-speed performance from zero to +0.8 V forward bias voltage. A 10 Gbit/sec error-free operation with a 20% power conversion efficiency is achieved. This result changes the common belief that high-speed PDs must be a power-consuming device under reverse bias.

Compared to III-V-based PDs, the major advantage of the Si-based PD is that it can be integrated with the mature CMOS ICs. Recently, IBM and Luxtera have demonstrated the integration of GOI waveguide PDs with CMOS ICs. It seems likely that PDs will play an important role in the fabrication of Si-based PICs in the near future. On the other hand, the III-V-semiconductor-based PDs and VCSELs have already become the major product in the OI market [47]. To further eliminate the power consumption of these well-developed high-speed PDs or even letting them generate DC electrical power and be monolithically integrated with 850 nm VCSELs [48] becomes an attractive way to further reduce the carbon footprint, package cost, and size of an OI system.

ACKNOWLEDGMENTS

The author thanks Professor J. E. Bowers and Molly Piels for their figures in the published journals and useful discussions.

REFERENCES

1. C. Lange, D. Kosiankowski, R. Weidmann, and A. Gladishch, "Energy consumption of telecommunication networks and related improvement options," *IEEE J. Sel. Top. Quantum Electron.*, vol. 17, pp. 285–295, 2011.
2. B. E. Lemoff, M. E. Ali, G. Panotopoulos, G. M. Flower, B. Madhavan, A. F. J. Levi, and D. W. Dolfi, "MAUI: Enabling fiber-to-processor with parallel multiwavelength optical interconnects," *IEEE/OSA J. Lightwave Technol.*, vol. 22, pp. 2043–2054, 2004.
3. L. Schares, J. A. Kash, F. E. Doany, C. L. Schow, C. Schuster, D. M. Kuchta, P. K. Pepeljugoski, et al., "Terabus: Terabit/second-class card-level optical interconnect technologies," *IEEE J. Sel. Top. Quantum Electron.*, vol. 12, pp. 1032–1044, 2006.
4. K. Kurata, "High-speed optical transceiver and systems for optical interconnects," *2010 Conference on OFC/NFOEC*, San Diego, CA, March 2010.
5. J. A. Kash, A. F. Benner, F. E. Doany, D. M. Kuchta, B. G. Lee, P. K. Pepeljugoski, L. Schares, C. L. Schow, M. Taubenblat, "Optical interconnects in exascale supercomputers," *IEEE Photonic Society Meeting 2010*, Denver, CO, November 2010.
6. R. S. Tucker, "A green Internet," *IEEE Lasers and Electro-Optics Society 2008 (LEOS/2008) Annual Meeting*, Newport Beach, CA, November 2008.
7. D. Bimberg, "Green data and computer communication," *IEEE Photonic Society Meeting 2011*, Arlington, VA, October 2011.
8. Clint L. Schow, "Power-efficient transceivers for high-bandwidth, short-reach interconnects," *Optical Fiber Communication Conference*, Los Angeles, CA, March 2012.
9. K. Yashiki, N. Suzuki, K. Fukatsu, T. Anan, H. Hatakeyama, and M. Tsuji, "1.1-μm-range high-speed tunnel junction vertical-cavity surface-emitting lasers," *IEEE Photon. Technol. Lett.*, vol. 19, pp. 1883–1885, 2007.
10. S. Imai, K. Takaki, S. Kamiya, H. Shimizu, J. Yoshida, Y. Kawakita, T. Takagi, et al., "Recorded low power dissipation in highly reliable 1060-nm VCSELs for "green" optical interconnection," *IEEE J. Sel. Topics Quantum Electron.*, vol. 17, no. 6, pp. 1614–1619, 2011.

11. J.-W. Shi, C.-C. Chen, Y.-S. Wu, S.-H. Guol, and Y.-J. Yang, "The influence of Zn-diffusion depth on the static and dynamic behaviors of Zn-diffusion high-speed vertical-cavity surface-emitting lasers at a 850 nm wavelength," *IEEE J. Quantum Electron.*, vol. 45, pp. 800–806, 2009.

12. J.-W. Shi, W.-C. Weng, F.-M. Kuo, J.-I. Chyi, S. Pinches, M. Geen, and A. Joel, "Oxide-relief vertical-cavity surface-emitting lasers with extremely high data-rate/power-dissipation ratios," *OFC/NFOEC*, Los Angeles, CA, 2011.

13. Y.-C. Chang, C. S. Wang, and L. A. Coldren, "High-efficiency, high-speed VCSELs with 35 Gbit/s error-free operation," *Electron. Lett.*, vol. 43, no. 19, pp. 1022–1023, 2007.

14. P. Moser, W. Hofmann, P. Wolf, J. A. Lott, G. Larisch, A. Payusov, N. N. Ledentsov, and D. Bimberg, "81 fJ/bit energy-to-data ratio of 850 nm vertical-cavity surface-emitting lasers for optical interconnects," *Appl. Phys. Lett.*, vol. 98, no. 23, pp. 231106, 2011.

15. C. T. DeRose, D. C. Trotter, W. A. Zortman, A. L. Starbuck, M. Fisher, M. R. Watts, and P. S. Davids, "Ultra-compact 45 GHz CMOS compatible germanium waveguide photodiode with low dark current," *Opt. Express,* vol. 19, no. 25, pp. 24897–24904, 2011.

16. L. Vivien, A. Polzer, D. Marris-Morin, J. Osmond, J. M. Hartmann, P. Crozat, E. Cassan, C. Kopp, H. Zimmermann, and J. M. Fedeli, "Zero-bias 40 Gbit/s germanium waveguide photodetector on silicon," *Opt. Express,* vol. 20, no. 2, pp. 1096–1101, 2012.

17. K. Wada, S. Park, and Y. Ishikawa, "Si photonics and fiber to the home," *IEEE Proc.*, vol. 97, no. 7, pp. 1329–1336, 2009.

18. L. Tsybeskov, D. J. Lockwood, M. Ichikawa, "Si photonics: CMOS going optical," *IEEE Proc.*, vol. 97, no. 7, pp. 1161–1165, 2009.

19. M. N. Sysak, H. Park, A. W. Fang, O. Raday, J. E. Bowers, and R. Jones, "Reduction of hybrid silicon laser thermal impedance using poly Si thermal shunts," *OFC/NFOEC*, Los Angeles, CA, March 2011.

20. A. Narasimha et al., "A 40-Gb/s QSFP optoelectronic transceiver in a 0.13 μm CMOS silicon-on-insulator technology," *Optical Fiber Communication Conference,* San Diego, CA, February 2008.

21. B. Yang, J. D. Schaub, S. M. Csutak, D. L. Rogers, and J. C. Campbell, "10-Gb/s all-silicon optical receiver," *IEEE Photon. Technol. Lett.*, vol. 15, pp. 745–747, 2003.

22. M. Jutzi, M. Berroth, G. Wohl, M. Oehme, and E. Kasper, "Ge-on-Si vertical incidence photodiodes with 39-GHz bandwidth," *IEEE Photon. Technol. Lett.*, vol. 17, pp. 1510–1512, 2005.

23. C. L. Schow, L. Schares, S. J. Koester, G. Dehlinger, R. John, and F. E. Doany, "A 15-Gb/s 2.4-V optical receiver using a Ge-on-SOI photodiode and a CMOS IC," *IEEE Photon. Technol. Lett.*, vol. 18, pp. 1981–1983, 2006.

24. G. Masini, G. Capellini, J. Witzens, and C. Gunn, "A four-channel 10 Gbps monolithic optical receiver in 130nm CMOS with integrated Ge waveguide photodetectors," *National Fiber Optic Engineers Conference*, Anaheim, CA, March 2007.

25. M. Morse, T. Yin, Y. Kang, O. Dosunmu, H. D. Liu, M. Paniccia, G. Sarid, et al., "State of the art Si-based receiver solutions for short reach applications," *Optical Fiber Communication Conference*, San Diego, CA, March 2009.

26. J. Schubert, E. Oliva, F. Dimroth, W. Guter, R. Loeckenhoff, and A. W. Bett, "High-voltage GaAs photovoltaic laser power converters," *IEEE Trans. Electron Devices,* vol. 56, pp. 170–175, 2009.

27. J.-W. Shi, F.-M. Kuo, Chan-Shan Yang, S.-S. Lo, and Ci-Ling Pan, "Dynamic analysis of cascade laser power converters for simultaneous high-speed data detection and optical-to-electrical DC power generation," *IEEE Trans. Electron Devices,* vol. 58, pp. 2049–2056, 2011.

28. M. Yang, K. Rim, D. L. Rogers, J. D. Schaub, J. J. Welser, D. M. Kuchta, D. C. Boyd, et al., "A high-speed, high-sensitivity silicon lateral trench photodetector," *IEEE Electron Device Lett.*, vol. 23, pp. 395–397, 2002.

29. M. Jutzi, M. Grözing, E. Gaugler, W. Mazioschek, and M. Berroth, "2-Gb/s CMOS optical integrated receiver with a spatially modulated photodetector," *IEEE Photon. Technol. Lett.*, vol. 17, pp. 1268–1270, 2005.

30. W.-K. Huang, Y.-C. Liu, and Y.-M. Hsin, "A high-speed and high-responsivity photodiode in standard CMOS technology," *IEEE Photon. Technol. Lett.*, vol. 19, pp. 197–199, 2007.

31. J.-W. Shi, F.-M. Kuo, F.-C. Hong, and Y.-S. Wu, "Dynamic analysis of a Si/SiGe based impact ionization avalanche transit time photodiode with an ultra-high gain-bandwidth product," *IEEE Electron Device Lett.*, vol. 30, pp. 1164–1166, 2009.

32. M.-J. Lee, H.-S. Kang, and W.-Y. Choi, "Equivalent circuit model for Si avalanche photodetectors fabricated in standard CMOS process," *IEEE Electron Device Lett.*, vol. 29, pp. 1115–1117, 2008.

33. W. S. Zaoui, H.-W. Chen, J. E. Bowers, Y. Kang, M. Morse, M. J. Paniccia, A. Pauchard, and J. C. Campbell, "Origin of the gain-bandwidth-product enhancement in separate-absorption-charge-multiplication Ge/Si avalanche photodiodes," *Optical Fiber Communication Conference*, San Diego, CA, March 2009.

34. S.-H. Huang, and W.-Z. Chen, "A 10-Gbps CMOS single chip optical receiver with 2-D meshed spatially-modulated light detector," *IEEE 2009 Custom Integrated Circuits Conference*, pp. 129–132, September 2009.

35. J. Oh, S. Csutak, and J. C. Campbell, "High-speed interdigitated Ge PIN photodetectors," *IEEE Photon. Technol. Lett.*, vol. 14, pp. 369–371, 2002.

36. H. Zang, S. J. Lee, W. Y. Loh, J. Wang, M. B. Yu, G. Q. Lo, D. L. Kwong, and B. J. Cho, "Application of dopant segregation to metal-germanium-metal photodetectors and its dark current suppression mechanism," *Appl. Phys. Lett.*, vol. 92, p. 051110, 2008.

37. H.-Y. Yu, S. Ren, W. S. Jung, A. K. Okyay, D. A. B. Miller, and K. C. Sarawat, "High-efficiency p-i-n photodetectors on selective-area-grown Ge for monolithic integration," *IEEE Electron Device Lett.*, vol. 30, pp. 1161–1163, 2009.

38. J. Oh, J. C. Campbell, S. G. Thomas, S. Bharatan, R. Thoma, C. Jasper, R. E. Jones, and T. E. Zirkle, "Interdigitated Ge p-i-n photodetectors fabricated on a Si substrate using graded SiGe buffer layers," *IEEE J. Quantum Electron.*, vol. 38, pp. 1238–1241, 2002.

39. M. Piels, A. Ramaswamy, W. Sfar Zaoui, J. E. Bowers, Y. Kang, and M. Morse, "Microwave nonlinearities in Ge/Si avalanche photodiodes having a gain bandwidth product of 300 GHz," *Conference on Optical Fiber Communication*, San Diego, CA, March 2009.

40. Y. Kang, H.-D. Liu, M. Morse, M. J. Paniccia, M. Zadka, S. Litski, G. Sarid, et al., "Monolithic germanium/silicon avalanche photodiodes with 340 GHz gain-bandwidth product," *Nature Photonics*, vol. 3, pp. 59–63, 2009.

41. M. Piels, and J. E. Bowers, "Si/Ge uni-traveling carrier photodetector," *Opt. Express,* vol. 20, no. 7, pp. 7488–7495, 2012.

42. H. Ito, S. Kodama, Y. Muramoto, T. Furuta, T. Nagatsuma, T. Ishibashi, "High-speed and high-output InP-InGaAs unitraveling-carrier photodiodes," *IEEE J. Sel. Top. Quantum Electron.*, vol. 10, pp. 709–727, 2004.

43. T. Yin, R. Cohen, M. M. Morse, G. Sarid, Y. Chetrit, D. Rubin, and M. J. Paniccia, "31GHz Ge n-i-p waveguide photodetectors on silicon-on-insulator substrate," *Opt. Express,* vol. 15, no. 21, pp. 13965–13971, 2007.

44. J.-W. Shi, C.-Y. Tsai, C.-S. Yang, F.-M. Kuo, Y.-M. Hsin, J. E. Bowers, and C.-L. Pan, "GaAs/In$_{0.5}$Ga$_{0.5}$P laser power converter with undercut mesa for simultaneous high-speed data detection and DC electrical power generation," *IEEE Electron Device Lett.*, vol. 33, pp. 561–563, 2012.

45. J.-W. Shi, Y.-T. Li, C.-L. Pan, M. L. Lin, Y. S. Wu, W. S. Liu, and J.-I. Chyi, "Bandwidth enhancement phenomenon of a high-speed GaAs-AlGaAs based unitraveling carrier photodiode with an optimally designed absorption layer at an 830 nm wavelength," *Appl. Phys. Lett,* vol. 89, p. 053512, 2006.

46. L. Zheng, X. Zhang, Y. Zeng, S. R. Tatavarti, S. P. Watkins, C. R. Bolognesi, S. Demiguel, and J. C. Campbell, "Demonstration of high-speed staggered lineup GaAsSb–InP unitraveling carrier photodiodes," *IEEE Photon. Technol. Lett.,* vol. 17, pp. 651–653, 2005.

47. J. A. Lott, A. S. Payusov, S. A. Blokhin, P. Moser, N. N. Ledentsov, and D. Bimberg, "Arrays of 850 nm photodiodes and vertical cavity surface emitting lasers for 25 to 40 Gbit/sec optical interconnects," *Phys. Status Solidi (C),* vol. 9, no. 2, pp. 290–293, 2012.

48. J.-W. Shi, F.-M. Kuo, T.-C. Hsu, Y.-J. Yang, A. Joel, M. Mattingley, and J.-I. Chyi, "The monolithic integration of GaAs/AlGaAs based uni-traveling-carrier photodiodes with Zn-diffusion vertical-cavity surface-emitting lasers with extremely high data-rate/power-consumption ratios," *IEEE Photon. Technol. Lett.,* vol. 21, pp. 1444–1446, 2009.

7 Quantum-Dot Nanophotonics for Photodetection

Ludan Huang and Lih Y. Lin

CONTENTS

7.1 INTRODUCTION

Traditionally, single-crystalline semiconductors materials have been employed as the active materials for photodetectors. Assisted by elaborative and sophisticated semiconductor fabrication tools, photodetectors utilizing these conventional materials have made great progress in the past few decades, enabling powerful technologies such as optical communication and digital imaging.

Colloidal quantum dots (QDs), as an alternative to the traditional active materials for photodetectors, have attracted much research and commercialization attention in recent years due to their unique properties. First, colloidal QDs can be flexibly integrated with a wide variety of substrates and functional components. These include complementary metal-oxide semiconductor (CMOS) silicon (Si) chips, flexible polymer substrates, and plasmonic enhancing structures. Second, they have great potential for low cost and large-scale production capability, as colloidal QDs can be synthesized from wet chemistry and processed into thin films using a variety of easy deposition methods, such as layer-by-layer self-assembly, drop casting, spin casting, and stamping. These deposition methods are performed under room temperature and ambient environment, eliminating the necessity for expensive high vacuum vapor deposition tools. Third, as an intrinsic nanometric material, it is a promising building block for achieving imaging/sensing arrays with ultrahigh resolution.

In this chapter, we review the work on photodetectors based on colloidal QDs in the light of these advantages. QD photodetectors fabricated by two representative methods—drop casting and electrostatic layer-by-layer self-assembly—will be reported. We will also give an example of QD photodetectors integrated with colloidal plasmonic enhancers—a completely solution-processed device. At the end, we will discuss the potential of using QD for high-resolution photodetector arrays.

The constant demand for smaller and faster computation and communication systems with versatile functionality has been driving the miniaturization, integration density and speed of device technologies. Although electronics have demonstrated tremendous success to date, photonics emerges as a promising alternative when it comes to the challenges that electronics face in speed, power consumption, and crosstalk between information channels. Like that for electronics, photonics and optoelectronics devices are traditionally based on solid-state semiconductor materials and achieved through elaborative top-down microfabrication methods. The continuous advancement of microfabrication technology has made possible novel nano- and microphotonic structures such as photonic crystals [1] and silicon slot waveguides [2]. One challenge that remains for photonic devices is diffraction limit and the resultant size mismatch between photonic and electronic component when it comes to hybrid integration.

On the other hand, the rapid advancement of synthesis of nanometric semiconductor materials in colloidal forms, or colloidal QDs, has provided another wealth of materials with unique optoelectronic properties and wet-chemistry-based fabrication capability. Various colloidal QD based nanophotonic devices, including waveguides [3,4], photovoltaics [5–7], and LED [8], as well as the integration between individual devices [9,10] have been proposed and studied.

Among all the photonic and optoelectronic devices, the photodetector is a critical component for both photonic–electronic integrated circuit and optical technologies, such as imaging, spectroscopy, and communication. In this chapter, we focus on the application of semiconductor colloidal QD in photodetection. The field has progressed rapidly in recent years and in-depth review papers are available [11]. In this chapter, rather than attempting to provide a comprehensive review of the field, we will use examples of the recent development on colloidal QD photodetectors in our group to illustrate the unique advantages of utilizing QD for photodetection, focusing on high spatial resolution, layer-by-layer self-assembly, and integration with plasmonics. The chapter is organized as follows: In Section 7.2, we introduce photodetectors fabricated by two different solution-processing methods—drop casting and electrostatic layer-by-layer self-assembly—then discuss the device performance. In Section 7.3, we give an example that demonstrates the flexibility of integrating a QD photodetector with other functional components. And in Section 7.4, we explore the prospect of making ultrahigh-resolution imaging and sensor arrays by studying the crosstalk in a QD photodetector array.

7.2 VERSATILE SOLUTION-BASED FABRICATION OF QUANTUM-DOT (QD) PHOTODETECTORS

Compared to conventional semiconductor thin films fabricated by vapor-deposition methods, colloidal semiconductor QDs feature the unique advantage of wet-chemistry batch synthesis and solution-based thin film formation. Through well-controlled batch synthesis, large quantity and highly monodispersed semiconductor QDs can be synthesized with versatile surface chemistry for functionalization, conjugation, and stabilization purposes. The solution-processing capability allows us to form uniform and large-area semiconductor thin films in a cost-effective manner. The QD thin film deposition techniques can be generally categorized into two types based on the solvents used for QD suspension. The first kind is usually processed from nonpolar organic solutions (such as chloroform, hexanes, toluene, etc.). During the processes, the QDs pack into multilayers while the solvent evaporates rapidly leaving only the solid thin films on the substrate. The bonding between QDs in such thin films is mainly Van der Waals force, so QD thin films will be redissolved when in contact with nonpolar solvents again. As a result, such QD thin films are mostly suitable for fabrication with no further postprocessing (e.g., photolithography) after the thin-film deposition. Drop casting, spin coating [12,13], dip coating [5], and inkjet printing belong to this category. The second kind is processed from aqueous solution in which QDs passivated with ionic functional groups (such as amine [–NH$_3^+$] or carboxylic acid [–COO$^-$]) form a colloidal suspension [14]. During the deposition process, the substrate modified with surface charges is dipped into the suspension of either negatively or positively charged QDs alternately. Each immersion results in a monolayer of QD deposition, and the procedure is repeated until the desired thickness is reached. The bonding between QDs in such thin films is the much stronger electrostatic Coulomb force and can therefore hold up to further postprocessing steps where various organic and inorganic solvents might be involved.

In terms of device configuration, QD photodetectors can be divided into lateral and vertical types. Lateral devices are usually photoconductive photodetectors as both metal contacts of the same material (and therefore the same work function) are placed in contact with the QD thin film horizontally [13,15]. External voltage is applied to the electrodes to collect the photoexcited electrons generated in the QD film. In contrast, the vertical devices usually employ contact material of different work functions to sandwich the QD thin film in between to form a photodiode structure [7,12,16]. The built-in electric field resulting from the work function difference sweeps the photoexcited carriers and generates photocurrent.

In the following subsections, we will describe two examples: a lateral QD photoconductive photodetector fabricated by drop casting and a vertical QD photodiode photodetector fabricated by electrostatic layer-by-layer self-assembly method. Their detailed fabrication and device performance will be discussed.

7.2.1 A LATERAL QD PHOTOCONDUCTIVE PHOTODETECTOR FABRICATED BY DROP CASTING

The photoconductive photodetector consists of a thin film of QDs that fills the gap of a lithographically defined metal nanojunction, as shown in Figure 7.1a. Ultrasmall

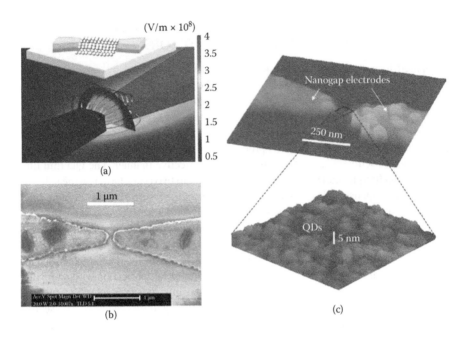

FIGURE 7.1 A nanoscale nanocrystal QD photodetector. (a) Illustration of the photodetector device structure and finite element model showing the normal DC electric field concentrated in the gap. Field lines and a cross-section of the field are shown. The unit of the scale bar is V/cm × 10^8. (b) SEM of a typical nanogap electrode without the nanocrystal thin film. (c) AFM of the device after QD deposition.

gap size in the range of ~25 to 50 nm (Figure 7.1b) is employed to create high filed intensities across the junction as well as to limit the number of tunneling steps required for electrons to traverse the gap, both of which increase the responsivity of the device [17].

Here we used commercially purchased CdSe/ZnS QDs passivated with long hydrocarbon chains (octadecylamine) and suspended in toluene. When carriers travel between QDs, the insulating ligands act as tunneling barriers and the tunneling probability exponentially decreases as the ligand length increases. Therefore, ligand removal is critical to obtaining QD thin films with high photoexcited carrier mobility. This is achieved via two steps: (1) partial ligand stripping in solution before the drop casting, and (2) thermal annealing the QD thin film in vacuum after the drop casting. The first step strips the organic capping ligand from the shell of the QDs by precipitating them out of the nonpolar host solvent with a polar organic reagent like ethanol or methanol. After the turbid QD suspension is centrifuged, the supernatant clear liquid is decanted and the QD precipitates are allowed to dry in a desiccator. The dried QDs can then be redispersed in a precast solvent of choice. The precast solvent plays an important role in determining the uniformity of the resultant thin film. For the results in this example, the commercially purchased CdSe/ZnS QDs originally suspended in toluene were washed twice and then resuspended in a 9:1 hexane–octane mixture.

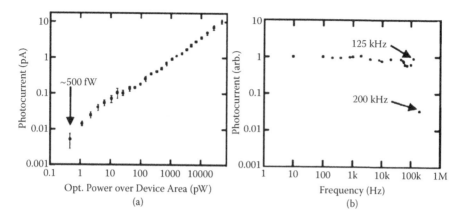

FIGURE 7.2 Experimental results for the nanocrystals QD photodetector. (a) Sensitivity measurement. An input optical power as low as 500 fW can be detected by the 50 nm nanogap QD photodetector. (b) Bandwidth measurement to 200 kHz. The result is limited by the bandwidth of the double lock-in experimental setup (125 kHz).

After micropipetting the precast QD solution onto an electrode-patterned substrate, the substrate is immediately transferred to a high vacuum chuck for thermal annealing at 400°C, which further removes the ligands by heating the device at above ligand vaporization temperature. Figure 7.1c shows an AFM (atomic force microscope) micrograph of the resultant thin film. The zoom-in shows that the QD thin film is quite uniform and smooth, with roughness on the order of the diameter of a single QD.

The fabricated QD photodetector devices were tested. The data showed that an optical power of lower than 500 fW can be detected by the device (Figure 7.2a). The device also demonstrated a bandwidth of at least 125 kHz (Figure 7.2b). This more than doubled the previously reported result for nanocrystal thin-film devices [13,18].

7.2.2 A Vertical QD Photodiode Photodetector Fabricated by Electrostatic Layer-by-Layer Self-Assembly

The photodiode photodetector consists of an intrinsic QD thin film that is sandwiched between top and bottom electrodes of different work functions [14]. As CdTe QDs, which are inherently capped with short ligands during the synthesis, are used to fabricate the QD thin film; no extra steps of ligand exchange, ligand removal, or annealing is needed to enhance the photoconductivity. Eliminating these steps is especially beneficial for integration with substrates that are sensitive to prolonged chemical treatment or vulnerable to high temperature annealing. In addition to shortening of interdot distance, surface passivation of QDs has shown beneficial effects on their photovoltaic device performance. It has been reported that thiol treatment can enhance external quantum efficiency by reducing mid-bandgap states, which serve as recombination centers and increase open circuit voltage by reducing metal-semiconductor junction interface states, which result in Fermi level pinning [6]. We use two kinds of thiol ligands for CdTe QDs synthesis in this work, 2-mercaptoethylamine (positively charged), and thioglycolic acid (negatively charged). Both kinds of molecules have one thiol-terminated end, which forms covalent bond with Cd and passivates the CdTe QD surface.

The fabrication process begins with silanizing the surface of a patterned indium tin oxide (ITO) substrate (Figure 7.3a) in order to create a positively charged surface for the deposition of the first monolayer of (negatively charged) QD self assembly. The silanization is done by first treating the substrate with oxygen plasma at 40 W for 10 min followed by an immediate 5 min immersion in 0.05 M NaOH to form hydroxyl groups on the surface. Then the chip is treated in (3-aminopropyl) triethoxysilane (APTES) solution (1 mL APTES in 20 mL toluene) at 70°C for 80 h, followed by sonication in MeOH/H$_2$O for 3 min to remove excess APTES on the surface. To protonate the amines, the chip is dipped in diluted HCl solution for a few seconds. Upon this step, there is high area density of positively charged amines on the ITO surface. The self-assembly is performed by alternatively pipetting negatively and positively charged QD solution onto the substrate, as shown in Figure 7.3b. Each self-assembly step is allowed an hour after which the chip is rinsed with a copious amount of deionized water and dried in vacuum. The entire process is performed in an N$_2$-filled glove box in the dark to avoid oxidation. For the device in this work, 65 layers of QDs were deposited. Subsequently, a thin layer of photoresist (AZ1512) was spin coated and two windows (1 × 1 mm^2) were opened by photolithography that defines the active area of the photodiode (Figure 7.3c). Finally, a thin layer of Al (around 50 nm) was deposited by e-beam evaporation. The final device structure is shown in Figure 7.3d.

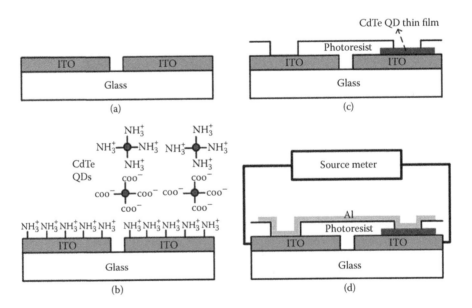

FIGURE 7.3   The overall fabrication process of the CdTe QD photodiode photodetector.

The structural illustrations and the absorbance/photoluminescence spectra of the lab synthesized positively charged MA-CdTe QDs and negatively charged TGA-CdTe QDs are shown in Figure 7.4. As previously mentioned, the extremely short ligand lengths of CdTe QDs, ~0.4 nm for MA, and ~0.35 nm for TGA enable the QD thin film to exhibit inherently superior photoconductivity. From the optical spectra, both MA- and TGA-CdTe QDs dispersed in water show photoluminescence peaks at 610 nm, which correspond to an estimated particle size (excluding capping ligand length) of around 3.5 nm.

The band structure of the photodiode under short-circuit condition (applied voltage = 0 V) is shown in Figure 7.5a. Under negative applied voltages, the photodiode shows open-circuit voltage of 0.4 V and short-circuit current of 43.2 μA/cm², with 1 mm × 1 mm active area under the illumination of a 405 nm laser with intensity of 53.5 mW/cm², as shown in Figure 7.5b. The short-circuit current corresponds to external quantum efficiency of 0.66%.

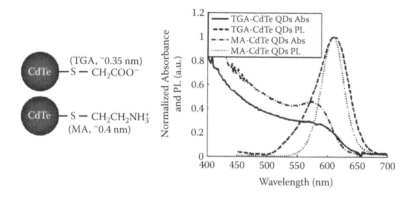

FIGURE 7.4 The structural illustrations and the absorbance/photoluminescence spectra of the CdTe QDs used for "photodiode" photodetectors.

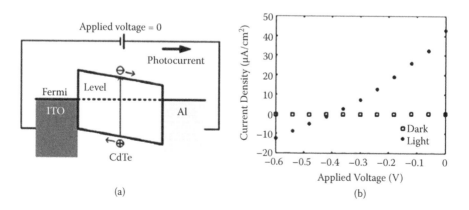

FIGURE 7.5 (a) The band structure of the CdTe QD photodiode under short-circuit condition. (b) The current density versus applied voltage measurement results.

7.3   PLASMON-ENHANCED QD PHOTODETECTORS

In the previous section, we discussed the processing versatility of using colloidal QDs for photodetection applications. In this section, we will present a plasmon-enhanced QD photodetector that further illustrates this idea. The plasmon-enhanced QD photodetector features the integration of colloidal plasmonic particles into colloidal QD-film photodetectors. This example is of particular interest because the all wet-chemistry-based integration of colloidal metal nanoparticles (NPs) and colloidal QD film proffers a promising candidate for future solutions of mass-production scalable, low-cost, light-absorbing optoelectronic devices [19].

7.3.1   DEVICE DESIGN AND FABRICATION

The device structure is based on the lateral nanoscale QD photodetector described in Section 7.2.1. The 3D top view and cross-section side view schematic drawings of the device are shown in Figure 7.6. An Ag NP layer is introduced between the QD layer and the glass substrate as the plasmonic enhancer. As the character of the fabrication approach adopted here is long-range uniformity rather than precise controlling of

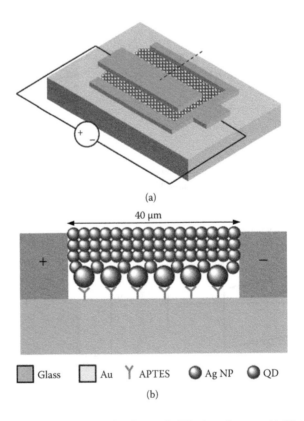

(a)

40 μm

Glass   Au   Y APTES   ● Ag NP   ● QD

(b)

FIGURE 7.6   Schematic drawings of a plasmonic QD photodetector. (a) 3D top view schematics and (b) cross-section view along the dash line in panel (a).

individual particle positions, we used 40 µm wide-gap electrodes to ensure that the measurement results represent large-scale average effect.

The fabrication of the device begins with patterning the glass substrate with comb electrodes using optical lithography followed by metallization and lift-off processes. The electrode-patterned substrate is then silanized with a monolayer of APTES, a process that is introduced in Section 7.2.2. After that, a monolayer of carboxylated Ag NPs, with a nominal diameter of 80 nm and an extinction peak at ~480 nm in solution, is self-assembled to the substrate. Using a high concentration colloidal solution (1 mg/mL) with relatively short assembly time (10 min) was critical to obtaining uniformly distributed high-density Ag NP monolayers. The optimized recipe yielded an average density of 42 particles/µm² on glass substrate. Figure 7.7a,b show the dark-field microscope images of the electrode-patterned substrate before and after the Ag NP deposition, respectively. It is noticed that the dark-field scattering of glass portion of the substrate changes from dark to purple after Ag NP deposition—a distinctive scattering feature of plasmon resonance in Ag NPs. An SEM (scanning electron microscope) micrograph of Ag NPs self-assembled on glass substrates is shown in Figure 7.7c. Finally, following the same procedure described in Section 1.1, partially ligand-removed CdSe/ZnS QDs with photoluminescence at 640 nm are drop cast onto the substrate.

7.3.2 Spectral Characterization and Modeling

Spectral characteristics of the plasmonic NPs, colloidal QDs, and the integrated plasmonic NPs-colloidal QDs composite are critical factors for device design as well as understanding of the device measurement results. The spectral position of the plasmon resonance is very sensitive to the change of dielectric environment and is therefore carefully characterized after each fabrication step.

In the plasmon enhanced QD photodetector devices, we used Ag NPs with a nominal diameter of 80 nm and an extinction peak at ~480 nm in solution as the plasmonic enhancer. We chose Ag NPs of relatively larger size because the absorption process dominates over the scattering process in smaller metal NPs [20], which is not desirable in plasmon enhancement applications. Compared to the extinction peak at ~480 nm in solution, the scattering resonance blue shifts to ~415 nm, which is expected considering the environmental refractive index change from water ($n = 1.33$) to mostly air ($n = 1$). This is also consistent with the purple color observed in the dark field image of Ag NPs modified glass substrate showed in Figure 7.7b.

After a QD drop cast, it is difficult to measure scattering of Ag NPs directly as they are embedded in the QD film. In this case, we measured the extinction of plasmonic NP-QD composite and compared it with the absorption of QD thin film of similar thickness as determined by a profilometer. Figure 7.8 shows the absorption enhancement spectrum, which is defined as the ratio of extinction of a plasmonic NP-QD composite layer to that of the QD only layer. A resonance at ~720 nm is shown, which indicates the plasmon resonance of Ag NPs when sandwiched between QD film and glass substrate. The kink at ~650 nm corresponds to the decrease of absorption in CdSe/ZnS QDs used in the devices. The CdSe/ZnS QDs have an emission

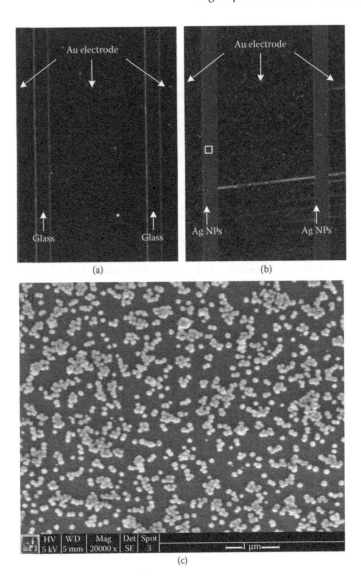

FIGURE 7.7 Plasmonic Ag NP deposition by self-assembly process. Dark field images of an electrode-patterned substrate (a) before and (b) after Ag NP deposition. After Ag NP deposition, the glass substrate changed from black to blue/purple color under microscope (not shown in black and white print). (c) A SEM micrograph of Ag NP self-assembled on glass substrate (position symbolically indicated in Figure 7.2b).

at ~650 nm with monotonically decreasing absorption from 400 nm to 600 nm, as shown in Figure 7.9.

Numerical simulations of the electromagnetic field distribution in device structures before and after QD deposition is performed by Lumerical FDTD Solutions simulation software package. The simulation results are summarized in Figure 7.10. It is clearly

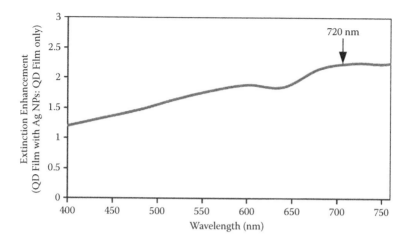

FIGURE 7.8 Extinction enhancement spectrum of QD film embedded with Ag NPs. Enhancement is defined as the ratio of extinction of Ag NP–QD composite layer to that of QD only layer.

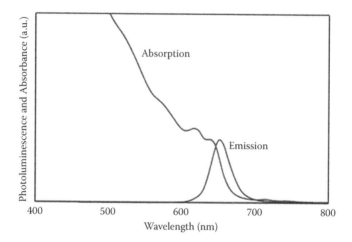

FIGURE 7.9 Photoluminescence and absorbance spectrum of CdSe/ZnS QDs.

shown that the plasmonic resonance (extinction cross-section) of Ag NPs is shifted from ~390 nm to ~750 nm after CdSe QD deposition, which is in good qualitative consistency with the experimental spectral characterization results (which show plasmon resonance at 415 nm and 720 nm before and after QD deposition, respectively). Enhanced absorption of QD film due to near-field and scattering enhancement from Ag NPs is the origin of wavelength-dependent responsivity enhancement of the device and, as will be demonstrated in the following section, indeed the trend of simulated scattering cross-section of Ag NPs (solid lightest gray line within the shaded region of the plot in Figure 7.10) is consistent with that of experimentally measured device responsivity enhancement.

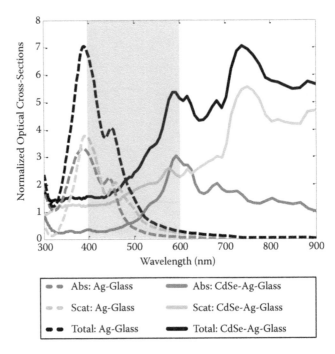

FIGURE 7.10 Dimensionless optical cross-section spectrum of 80 nm diameter Ag NPs on glass substrate before (dashed lines) and after (solid lines) CdSe QD deposition. The dimensionless optical cross-sections are calculated as the ratios of simulated optical cross-section areas to the geometrical cross-section area of Ag NPs (i.e., πr^2). The shaded area indicates the wavelength range in which device photocurrent measurements were taken. (From L. Huang, C. C. Tu, and L. Y. Lin, *Applied Physics Letters*, 98, 113110, 2011 [19].)

7.3.3 RESULTS

To investigate how Ag NPs influence the performance of QD photodetectors, photocurrents generated at different wavelengths of illumination was measured, and the result was compared with QD photodetectors processed from the same QD solution on the same substrate but without Ag NP integration. It is worth mentioning that careful comparison of dark currents was made between devices with Ag NPs and that without Ag NPs before QD deposition as well as after QD deposition. No consistent difference was measured in both cases, which indicated that the Ag NPs did not cause noticeable leakage current at present particle density level.

Device responsivity was calculated based on the average photocurrent and the power of light incident upon the active area of the device. Figure 7.11a,b each shows the responsivity of two devices of similar film thicknesses, one with Ag NPs and one without, over the spectral range from 400 nm to 600 nm. By using a profilometer, the overall film thicknesses are determined to be ~440 nm for both devices in Figure 7.11a and ~100 nm for both devices in Figure 7.11b. It is evident that for both film thicknesses the responsivity of the device with Ag NPs was higher than that without Ag NPs over the entire measured spectral range.

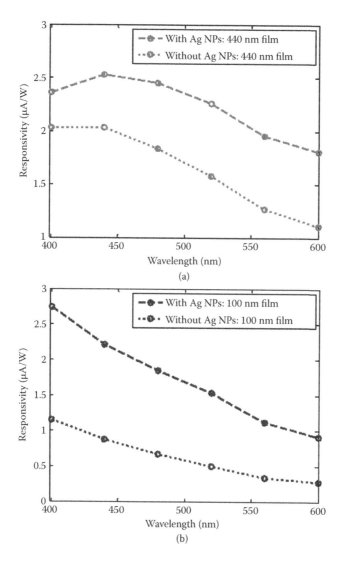

FIGURE 7.11 Device responsivity measurement results. (a) Responsivity for two QD photodetectors of 440 nm film thickness with (dashed line) and without (dash-dot line) Ag NPs. (b) Responsivity for two QD photodetectors of 100 nm film thickness with (dashed line) and without (dash-dot line) Ag NPs. Devices were biased at 20 V for responsivity measurement.

Responsivity enhancement, which is defined as the ratio of responsivity of a device with Ag NPs to one without Ag NPs, is 1.2- to 1.6-fold enhancement for the 440 nm thick QD film device and a 2.4- to 3.3-fold enhancement for the 100 nm thick QD device, as shown in Figure 7.12. It is noticed that the responsivity enhancement increases with wavelengths. This phenomenon is expected due to two factors. First, as indicated by both film absorption measurement (Figure 7.8) and simulation (Figure 7.10), the plasmon resonance of Ag NPs shifts to above 700 nm region

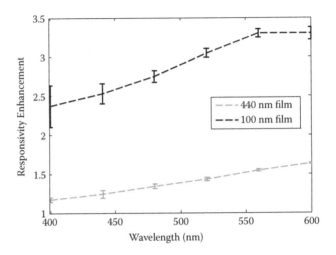

FIGURE 7.12 Device responsivity enhancement results. Responsivity enhancement for devices with 440 nm (gray dashed line) and 100 nm (black dashed line) film thicknesses.

(with the exact spectral positions to be 720 nm experimentally and 750 nm theoretically) after the QD deposition due to the dielectric constant change of the environment materials. As a result, the longer wavelength part of the measured spectrum lies closer to plasmon resonance and is enhanced more. Second, the wavelength-dependent absorption of QD film allows more longer-wavelength portions of the incident spectrum to reach the plasmonic particles, and therefore relatively increases the input to plasmonic components at the longer wavelength. This effect of QD-film altered input spectrum for plasmon NPs is further confirmed when we measured the enhancement of a set of devices with thinner QD films. As shown in Figure 7.12, the responsivity enhancement for a 100 nm QD film (black dashed line) is higher across the spectrum as the thinner QD film allows more input power to transmit to plasmon particles.

7.4 CROSSTALK IN A HIGH-DENSITY QD PHOTODETECTOR ARRAY

In this section, we will look at one potential advantage of nanometric-gap QD photodetectors that has not been extensively studied yet: their prospect to make ultrahigh resolution imaging and sensor arrays. Hegg et al. [17] demonstrated by modeling that the E field of a nanometric-gap electrode is highly concentrated in the gap, as shown in Figure 7.13a, resulting in an ultrasmall device active area. This suggests possible low crosstalk even in an ultrahigh density array. Here we investigate crosstalk in closely positioned QD photodetectors.

7.4.1 MEASUREMENT SCHEME

Crosstalk, in general, is any phenomenon by which a signal transmitted on one element creates an undesirable effect in an adjacent element. In an image sensor

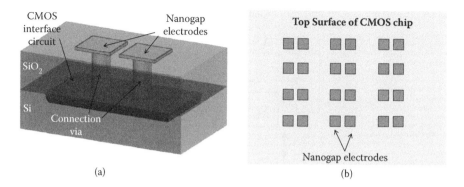

FIGURE 7.13 (a) 3D schematics of photodetector electrode integrated with CMOS circuitry. (b) A top view schematics of nanogap photodetector array. (From L. Huang, M. Strathman, and L. Y. Lin, *Optics Letters*, 37(15), 314–3146, 2012 [24].)

array, it can be specified as photocurrent induced in adjacent pixels when a central pixel is illuminated.

$$\text{Crosstalk} = \frac{\text{Signal of adjacent (unexposed) pixels}}{\text{Signal of the central (exposed) pixel}} \tag{7.1}$$

In current CMOS image arrays, the effect of crosstalk is usually investigated by methods such as optical shielding of adjacent pixels [21] or scanning a micron-size focused spot over an array of pixels [22,23] while measuring the response of adjacent pixel and compare it to that of the central pixel. Although these methods are intuitive and direct to implement, they rely on the fact that the pixel size of the investigated image sensor array being at least a few microns in size and light can be confined down to such scale without much difficulty imposed by diffraction.

QD photodetectors, on the other hand, proffers the potential to make pixel size much smaller than the state-of-art CMOS image sensor technology, down to approaching or even beyond diffraction limit of light. One of the reasons that QD photodetectors have potential for big resolution advancement compared to a CMOS image array is that colloidal QDs can be easily integrated onto the top of CMOS circuitry, as shown in Figure 7.13. In contrast to a conventional CMOS image array, where each pixel consists of an area for photosensing and area for control/read-out circuit separately, in a QD photodetector–CMOS chip integrated unit each pixel only consumes area for photosensing, as the photosensing element and the control/read-out circuit element are on different layers of the structure, connecting by via holes. The separation of photosensing area and control/read-out circuit area alone has the potential to significantly increase resolution.

Because the conventional methods are no longer applicable as the pixel size approaching the wavelength of incident light, we developed a new approach to evaluate the crosstalk effect in high-density QD nanometric-gap array. Instead of confining the illumination to an individual pixel, the entire sensor array is uniformly illuminated with incident light. The photocurrent is monitored in a constantly

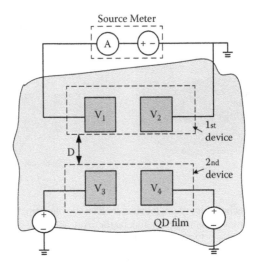

FIGURE 7.14 Experiment scheme for photodetector resolution measurement. The entire field is covered with QD and illuminated by a 405 nm laser. (From L. Huang, M. Strathman, and L. Y. Lin, *Optics Letters*, 37(15), 314–3146, 2012 [24].)

activated device, while the adjacent device is switched on and off during the measurement, as shown in Figure 7.14. For simplicity, the two devices are referred to as the "1st device" and the "2nd device" onward. The crosstalk effect is then evaluated by the amount of change in photocurrent induced in the 1st device when the 2nd device is switched between the ON and OFF states, as in Equation (7.2):

$$\text{Crosstalk} = \frac{\left| \text{Signal (1stdevice)}_{\text{2nddeviceON}} - \text{Signal (1stdevice)}_{\text{2nddeviceOFF}} \right|}{\text{Signal (1stdevice)}_{\text{2nddeviceOFF}}} \quad (7.2)$$

The proposed scheme measures two types of crosstalk: optical crosstalk and electrical crosstalk. Optical crosstalk occurs when two active devices are positioned close enough that the active area of the devices overlaps, as shown in Figure 7.15a. Optical crosstalk manifests as a decrease in photocurrent in the 1st device when the 2nd device is ON compared to when the 2nd device is OFF, as when active the 2nd device would draw photocurrent from the overlapped active area. Electrical crosstalk, on the other hand, arises from modified current paths resulted from different bias assignment to electrodes when the 2nd device is switched between ON and OFF. An example of this effect is illustrated in Figure 7.15b. Here electron paths to electrode 1 (E1) of two bias configurations are compared (note that the source meter reads only charges collected by E1). The first bias configuration is (V1, V2, V3, V4) = (5 V,0 V,5 V,5 V) (noted as (5055) onward and labeled with a black arrow in Figure 7.15b) and the second bias configuration is (5050) (labeled with gray arrows in Figure 7.15b). It is obvious that with configuration (5050), which corresponds to the 2nd device ON, an additional electron path from E4 to E1 is introduced. It results in higher photocurrent in the 1st device when the 2nd is ON comparing to when the

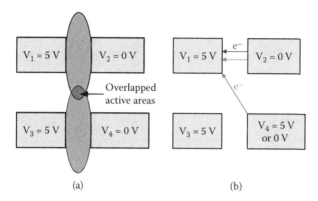

(a)   (b)

FIGURE 7.15 Schematic drawings of different mechanisms of crosstalk. (a) Optical crosstalk. (b) Electrical crosstalk.

2nd device is OFF. It is important to notice that, when the bias configurations are chosen properly (as the example given in Figure 7.15), optical crosstalk and electrical crosstalk have opposite signs.

7.4.2 Plain Planar Electrode Devices

To investigate the crosstalk effect in QD photodetector/sensor arrays, planar devices separated with different distances are fabricated. Figure 7.16a shows the resulting closely positioned electrodes. Figure 7.16b is a close-up image of the central nanogap area of the devices. The nanogap is 55 nm and the spacing between adjacent electrodes is 212 nm. Following procedures described in Section 7.2.1, QDs are deposited onto the SiO_2/Si substrate. Figure 7.17 shows the electrodes covered by a uniform QD thin film after the drop cast and thermal annealing steps.

We first studied the photocurrent characteristics at different bias configurations and the result is summarized in Figure 7.18. The photocurrents at (5,0,0,0) and (5,0,0,5) are several times larger than those at (5,0,5,5) and (5,0,5,0), suggesting the leakage current from E3 to E1 is a main contribution to photocurrent reading when E3 is biased LOW (0 V). This is understandable because although the charge tunneling distance is larger between E3 and E1 than that between E2 and E1, the charge generation cross-section is much bigger, which results in a large leakage current. This finding also suggests that in high-density QD photodetector arrays, it is preferable to bias the idle devices HIGH to alleviate leakage current.

To eliminate leakage current between E3 and E1, in the following testings we biased the 2nd device at (V3,V4) = (5,5) for the OFF state and at (V3,V4) = (5, 0) for the ON state. Figure 7.19 shows the response of the 1st device under different bias configurations in time sequence. The results are summarized in Table 7.1 and an average change of 25.8% in photocurrent is measured. Note that the photocurrent is higher with the 2nd device turned ON, suggesting that an addition photocurrent between E1 and E4 (i.e., electrical crosstalk) is a more significant source than the overlapping optical active area (i.e., optical crosstalk).

(a)   (b)

FIGURE 7.16   E-beam patterned electrodes for two devices positioned at 212 nm separation. (a) A low magnification SEM image. Scale bar is 200 μm. (b) A close-up SEM image for the central nanogap area. Scale bar is 500 nm. (From L. Huang, M. Strathman, and L. Y. Lin, *Optics Letters*, 37(15), 314–3146, 2012 [24].)

FIGURE 7.17   The same electrodes as shown in Figure 7.16 after QD deposition. (From L. Huang, M. Strathman, and L. Y. Lin, *Optics Letters*, 37(15), 314–3146, 2012 [24].)

In the aforementioned design of closely positioned QD photodetectors, the electrodes of the two devices are parallel and elongated in one direction for probing purpose (see Figure 7.16b). This long parallel electrode configuration is a significant source of leakage current and crosstalk in the experiment, as there is a large quantity of QDs between the elongated electrodes (high field region during device operation). In future applications of imaging/sensing devices fabricated with CMOS processing, the QD detectors are expected to be integrated with the underlying CMOS circuitry

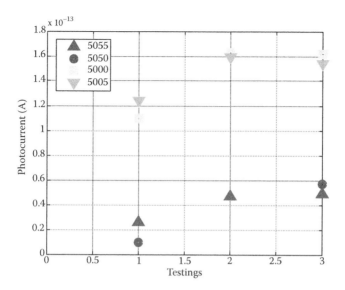

FIGURE 7.18 Photocurrent from the 1st device under various bias configurations.

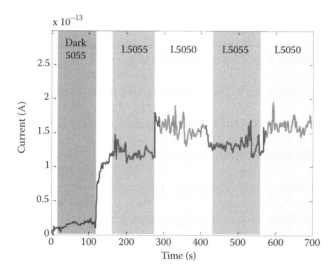

FIGURE 7.19 Photocurrent from the 1st device under various bias configurations. As an example for labeling, "L5000" represents devices under illumination with bias configuration of (V1, V2, V3, V4) = (5 V, 0 V, 5 V, 5 V). D = 212 nm. (From L. Huang, M. Strathman, and L. Y. Lin, *Optics Letters*, 37(15), 314–3146, 2012 [24].)

TABLE 7.1

Photocurrent and Crosstalk from the Device Tested in Figure 7.19 [60]

Testing Cycles	Photocurrent at 5055 (A)	Photocurrent at 5050 (A)	Crosstalk
1	1.01e–13±0.06e–13	1.33e–13±0.11e–13	32.4±12.5%
2	1.18e–13±0.12e–13	1.41e–13±0.07e–13	19.3±12.6%

Source: From L. Huang, M. Strathman, and L. Y. Lin, *Optics Letters*, 37(15), 314–3146, 2012 [24].

through via holes, as shown in Figure 7.13a. The smaller electrode areas in the top layer (Figure 7.13b) will substantially minimize the aforementioned problem.

7.4.3 Masked Planar Electrode Devices

In the interest of obtaining a closer estimation to the crosstalk of QD photodetectors/sensors integrated onto a CMOS chip as shown in Figure 7.13, we modified the electrode structures. A main feature of a QD photodetector/sensor array integrated with CMOS circuit is the small area of electrodes on the top layer. This is possible because in CMOS processing multiple layers of metals and dielectrics are allowed, and the signal is read out by metal wiring on an underneath layer, which connects to the top electrodes through vias. Although we do not have readily accessible fabrication capability to make multiple metal layers, we designed a masked planar electrode layout that enable nanoscale electrodes and millimeter-scale read-out pads on one metal layer. Figure 7.20 depicts the fabrication procedures of the masked planar electrode devices. Similar to plain planar devices, the fabrication process starts with E-beam patterning electrodes on 1 μm SiO$_2$/Si wafers, followed by metallization and lift-off, as shown in Figure 7.20a,b. After that, instead of depositing QD directly, a Si$_3$N$_4$ layer of 400 nm is deposited by plasma-enhanced chemical vapor deposition, as shown in Figure 7.20c. Subsequently, a second E-beam lithography is performed to pattern a central window area and four pad areas. After developing the pattern, a reactive ion etching is performed to etch the exposed Si$_3$N$_4$ regions using PMMA as a mask. Figure 7.20d,e show the resultant 3D and top view of the Si$_3$N$_4$-masked planar electrode structure. The central window defines the area for electrodes, while the four pad areas provide the access to probe the devices. After this, QDs prepared using the same method are drop cast onto the wafer, as shown in Figure 7.20f. By adopting this fabrication procedure, we successfully created photodetector electrodes with sizes similar to those in a future integration with CMOS using just one layer of metal. SEM images of the electrodes at different steps are shown in Figure 7.21.

Figure 7.22 shows the response of the 1st device under different bias configurations in time sequence, with the two devices separating at 200 nm. As the amount of QDs around the electrodes was much less than it was in the case of a plain planar electrode device array, a higher voltage of 20 V was applied as HIGH to collect measureable current. Comparing to the results of previous plain planar electrode devices, it is noticed that the current at (5000) (note that E3 is biased at LOW) is comparable to the current at

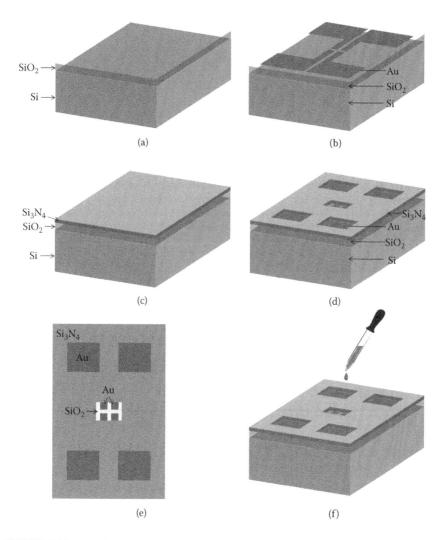

FIGURE 7.20 Fabrication process of a Si_3N_4-masked planar electrode device.

(5055) and (5050) instead of being a few times larger. This confirmed that the masked planar electrode design did decrease electrical crosstalk significantly. The crosstalk results with E3 biased at HIGH are summarized in Table 7.2 and an average change of 8.4% in photocurrent is measured. Notice that as expected the value is significantly lower than the crosstalk for plain planar electrode devices (25.8%).

From the crosstalk measurement, we can also obtain some further understanding of the active device area for nanometric gap QD photodetectors. It was assumed that the active area of a single device is confined to the nanometric gap area based on finite element modeling results of the E field distribution [17]. However the crosstalk measurement showed that the device drew current from a larger area though the exact size is not determined from the present measurements. This is because

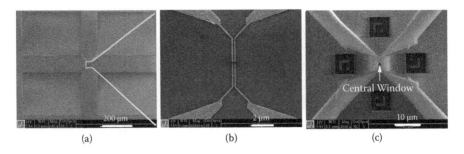

(a)   (b)   (c)

FIGURE 7.21 SEM images of electrodes at different steps prior to QD deposition. (a) An overall image of electrodes and (b) a close-up image of central electrode region at the step illustrated in Figure 7.20b. (c) A close-up image of central electrode region at the step illustrated in Figure 7.20d,e. The red arrow points out the central window region. Note that the magnification in (b) and (c) are different.

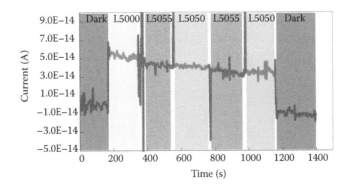

FIGURE 7.22 Photocurrent from the 1st device under various bias configurations. "5" represents 20 V. For example, "L5000" represent devices under illumination with bias configuration of (V1, V2, V3, V4) = (20 V, 0 V, 20 V, 20 V). D = 200 nm.

TABLE 7.2
Photocurrent and Crosstalk from the Device Tested in Figure 7.22

Testing Cycles	Photocurrent at 5055 (A)	Photocurrent at 5050 (A)	Crosstalk
1	4.19e–14 ± 0.27e–14	4.05e–14 ± 0.27e–14	3.52 ± 8.97%
2	4.60e–14 ± 0.24e–14	5.21e–14 ± 0.22e–14	13.30 ± 7.66%

although the E field in the other parts of the electrode is more than two orders of magnitude lower than that in the nanogap, the current generation cross-section is much larger. Furthermore, when QD photodetector devices are made into arrays, the device active area is additionally affected by the layout of electrodes and devices as well as bias configuration.

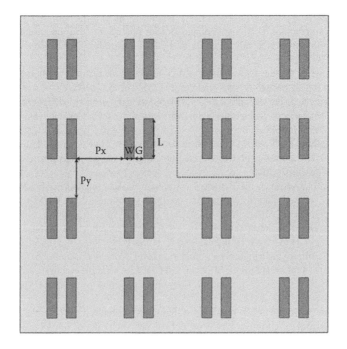

FIGURE 7.23   Top-view schematics of electrodes for a QD nanometric photodetector array.

It is of interest to assess the resolution or density improvement that can be achieved by a QD photodetector array compared to existing technology. The average crosstalk of 8.4% ($<e^{-2}$) would suggest that device separation of 200 nm is well acceptable. However, considering that there are four immediate adjacent devices in an array (and it should be noted that the two adjacent devices in horizontal direction might contribute more than those in vertical direction depending on the bias configuration), we assume the following parameters (denoted in Figure 7.23) in a QD array: W = G = 100 nm, Px = 500 nm, Py = L = 400 nm, which gives a pixel size of 800 nm × 800 nm. Compared to the state-of-art 1.75 μm pitch CMOS imager, the resolution or density improvement is 4.8-fold! The actual improvement may be more or less depending on the aforementioned array effect, but the potential of revolutionary resolution improvement is evident.

In summary, in this chapter we reviewed our work on photodetectors based on colloidal QDs. The work ranges from individual devices, to device integration, to photodetector sensor array. At all the different levels, QD photodetectors demonstrate fabrication and integration versatility through solution-based processing as well as proven advantages for further device/system miniaturization compared to the traditional bulk semiconductor technologies.

REFERENCES

1. S. G. Johnson, P. R. Villeneuve, S. Fan, and J. D. Joannopoulos, "Linear waveguides in photonic-crystal slabs," *Phys. Rev. B,* 62, 8212–8222, 2000.

2. W. Bogaerts, R. Baets, P. Dumon, V. Wiaux, S. Beckx, D. Taillaert, B. Luyssaert, J. Van Campenhout, P. Bienstman, and D. Van Thourhout, "Nanophotonic waveguides in silicon-on-insulator fabricated with CMOS technology," *J. Lightwave Technol.*, 23, 401–412, 2005.

3. C. J. Wang, L. Huang, B. A. Parviz, and L. Y. Lin, "Sub-diffraction photon guidance by quantum dot cascades," *Nano Letters*, 6(11), 2549–2553, 2006.

4. C. J. Wang, B. A. Parviz, and L. Y. Lin, "Two-dimensional array self-assembled quantum dot sub-diffraction waveguides with low loss and low crosstalk," *Nanotechnology*, 19(23), 295201, 2008.

5. J. M. Luther, M. Law, M. C. Beard, Q. Song, M. O. Reese, R. J. Ellingson, and A. J. Nozik, "Schottky solar cells based on colloidal nanocrystals films," *Nano Letters*, 8(10), 3488–3492, 2008.

6. D. A. R. Barkhouse, A. G. Pattantyus-Abraham, L. Levina, and E. H. Sargent, "Thiols passivate recombination centers in colloidal quantum dots leading to enhanced photovoltaic device efficiency," *ACS Nano*, 2(11), 2356–2362, 2008.

7. X. Wang, G. I. Koleilat, J. Tang, H. Liu, I. J. Kramer, R. Debnath, L. Brzozowski, D. A. R. Barkhouse, L. Levina, S. Hoogland, and E. H. Sargent, "Tandem colloidal quantum dot solar cells employing a graded recombination layer," *Nature Photonics*, 5, 480–484, 2011.

8. C. C. Tu, L. Tang, J. Huang, A. Voutsas, and L. Y. Lin, "Visible electroluminescence from hybrid colloidal silicon quantum dot-organic light-emitting diodes," *Applied Physics Letters*, 98, 213102, 2011.

9. M. Naruse, T. Miyazaki, T. Kawazoe, S. Sangu, K. Kobayashi, F. Kubota, and M. Ohtsu, "Nanophotonic computing based on optical near-field interactions between quantum dots," *IEICE Transactions on Electronics*, E88C(9), 1817–1823, 2005.

10. L. Y. Lin, C. J. Wang, M. C. Hegg, and L. Huang, "Quantum dot nanophotonics—From waveguiding to integration," *Journal of Nanophotonics*, 3(1), 031603, 2009.

11. G. Konstantatos, and E. H. Sargent, "Nanostructured materials for photon detection," *Nature Nanotechnology*, 5, 391–400, 2010.

12. D. C. Oertel, M. G. Bawendi, A. C. Arango, and V. Bulović, "Photodetectors based on treated CdSe quantum-dot films," *Applied Physics Letters*, 87, 213505, 2005.

13. G. Konstantatos, I. Howard, A. Fischer, S. Hoogland, J. Clifford, E. Klem, L. Levina, and E. H. Sargent, "Ultrasensitive solution-cast quantum dot photodetectors," *Nature*, 442, 180–183, 2006.

14. C.-C. Tu, and L. Y. Lin, "Thin film photodiodes fabricated by electrostatic self-assembly of aqueous colloidal quantum dots," *Thin Solid Films*, 519(2), 857–862, 2010.

15. G. Konstantatos, L. Levina, J. Tang, and E. H. Sargent, "Sensitive solution-processed Bi_2S_3 nanocrystalline photodetectors," *Nano Letters*, 8(11), 4002–4006, 2008.

16. J. P. Clifford, G. Konstantatos, K. W. Johnston, S. Hoogland, L. Levina, and E. H. Sargent, "Fast, sensitive and spectrally tuneable colloidal-quantum-dot photodetectors," *Nature Nanotechnology*, 4, 40–44, 2009.

17. M. C. Hegg, M. P. Horning, T. Baehr-Jones, M. Hochberg, and L. Y. Lin, "Nanogap quantum dot photodetectors with high sensitivity and bandwidth," *Applied Physics Letters*, 96, 101118, 2010.

18. V. J. Porter, S. Geyer, E. Halpert, M. Kastner, and M. Bawendi, "Photoconduction in annealed and chemically treated CdSe/ZnS inorganic nanocrystal films," *Journal of Physical Chemistry C*, 112, 2308–2316, 2008.

19. L. Huang, C. C. Tu, and L. Y. Lin, "Colloidal quantum dot photodetectors enhanced by self-assembled plasmonic nanoparticles," *Applied Physics Letters*, 98, 113110, 2011.

20. P. K. Jain, K. S. Lee, I. H. El-Sayed, and M. A. El-Sayed, "Calculated absorption and scattering properties of gold nanoparticles of different size, shape, and composition: Applications in biological imaging and biomedicine," *Journal of Physical Chemistry B*, 110(14), 7238–7248, 2006.

21. C.-C. Wang, and C. G. Sodini, "A crosstalk study on CMOS active pixel sensor arrays for color imager applications," http://www.imagesensors.org/Past%20Workshops/2001%20Workshop/2001%20Papers/pg%20068%20CWang.pdf.
22. G. Agranov, V. Berezin, and R. H. Tsai, "Crosstalk and microlens study in a color CMOS image sensor," *IEEE Transactions on Electron Devices*, 50(1), 4–11, 2003.
23. T. J. Martin, M. J. Cohen, J. C. Dries, and M. J. Lange, "InGaAs/InP focal plane arrays for visible light imaging," http://lib.semi.ac.cn:8080/tsh/dzzy/wsqk/SPIE/vol5406/5406-38.pdf.
24. L. Huang, M. Strathman, and L. Y. Lin, "Exploring spatial resolution in high-sensitivity nanogap quantum dot photodetectors," *Optics Letters*, 37(15), 314–316, 2012.

8 Rolled-Up Semiconductor Tube Optical Cavities

Pablo Bianucci, M. Hadi Tavakoli Dastjerdi, Mehrdad Djavid, and Zetian Mi

CONTENTS

8.1   INTRODUCTION

Optical microcavities are structures wherein light can be confined at microscopic scales. Due to the strong confinement, the optical density of states of the cavity develops resonance peaks at discrete wavelengths. The corresponding resonant modes of the cavity can be significantly enhanced with respect to the same electromagnetic field in free space [1]. As a consequence, the interaction between the strongly confined photons and matter changes drastically within these cavities. These changes can range from reduced spontaneous emission lifetimes (known as the Purcell effect) [2,3] in the weak coupling regime to the appearance of Rabi splitting signaling the strong coupling regime between the electromagnetic field inside the resonator and a quantum system [4]. The magnitude of these effects depends on the strength of the optical confinement of the cavity modes and their spatial extent. Semiconductor materials are attractive for microcavities for a number of reasons. The high refractive index of semiconductors allows for the design of compact microcavities with strong optical confinement. In addition, high-efficiency quantum emitters, such as quantum dots, wires, and wells can be readily grown and engineered by modern epitaxial growth techniques including molecular beam epitaxy (MBE) [5], chemical beam epitaxy (CBE) [6] and metalorganic chemical vapor phase deposition (MOCVD) [7]. Furthermore, semiconductors offer the possibility of electrical injection of carriers into the devices [8], a very desirable attribute for the fabrication of practical nano-optoelectronic devices. In this context, a variety of microcavity designs have been intensively investigated, such as microdisks [9] and microrings [10], micropillars [11], and photonic crystals [12], to name just a few. In disks and pillars, in-plane confinement is obtained by the total internal reflection caused by the high semiconductor refractive index, whereas in photonic crystals this confinement is the product of Bragg reflections in a two-dimensional periodic lattice. Micropillars keep light confined in the vertical direction by using one-dimensional Bragg reflectors to form a Fabry-Perot cavity. The vertical confinement in microdisks and photonic crystals labs is provided again by the total internal reflection related to the steep refractive index difference between the resonator material and the surrounding medium. More recently, even smaller cavity structures have been developed, such as nanoneedles [13] and nanoplasmonic [14] cavities.

Another recent addition to the semiconductor optical microcavity repertoire is the rolled-up microtube [15]. These tube cavities are formed by the self-rolling of strained nanomembranes as they are released from the host substrates. They can be manufactured in various material systems that allow for the creation and subsequently selective release of strained membranes. Combining the advantages of both top-down and bottom-up fabrication processes, such rolled-up tube cavities offer an unprecedented control over the confined optical modes [16,17] and the emission characteristics [18,19]. These properties, in conjunction with the possibility of transferring them onto foreign substrates [20] set them apart from the previously discussed microcavities. In this regard, rolled-up semiconductor tubes have emerged as promising candidates for applications in optical communications, biological sensing, and micro- and nanofluidics. Proof-of-concept demonstrations, such as optically pumped lasers [21,22] and optofluidic sensors [23] have already been reported.

In this chapter, we will first discuss the fabrication of rolled-up semiconductor tube optical cavities in Section 8.2, with a special focus on the fabrication of GaAs and InP-based tube structures. This is followed by a brief discussion of the methods used to transfer microtubes onto silicon and other foreign substrates in Section 8.3. In Section 8.4, we will present a detailed description of their optical properties, including both modeling and experimental results. The recent achievement of rolled-up microtube lasers is discussed in Section 8.5. We will also briefly address some of the emerging applications of rolled-up semiconductor tubes in Section 8.6. Finally, concluding remarks are made in Section 8.7.

8.2 FABRICATION OF ROLLED-UP SEMICONDUCTOR TUBES

To date, rolled-up semiconductor tubes have been demonstrated using a variety of materials, such as InGaAs/GaAs [21], InGaAsP [24], SiO$_x$/Si [25], SiGe [26], and AlN/GaN [27]. The tube diameters can be varied from hundreds of nanometers to tens of micrometers. Additionally, rolled-up tubes can be made of metals and polymers [28,29]. In what follows, the fabrication and formation mechanism of semiconductor tubes is first described, followed by a brief overview of rolled-up metal and polymer tube structures.

The formation of rolled-up microtubes, first discovered by Prinz and collaborators in 2000 in the GaAs/InGaAs system [30], is based on the release of strain in nanomembranes. Although we will focus on this system to illustrate the fabrication process, the same approach is, in principle, applicable to any other systems wherein strained nanomembranes can be formed. Figure 8.1 shows a schematic of an InAs/GaAs membrane and the related tube formation process. GaAs has a smaller lattice constant than InAs. As a result, the GaAs layer is tensile strained after its pseudomorphic deposition on top of the InAs. Rolled-up tubes are formed because of the strain relaxation in the membrane when it is released from the substrate. As a consequence, the curvature of the rolled-up tube can be controlled by adjusting the built-in strain in the bilayer and the respective thicknesses of its constituent membranes. With the use of a continuum mechanics model [31], it is possible to predict the diameter of the resulting microtube with the following formula:

$$D = \frac{1}{3} \frac{1}{\varepsilon} \frac{(d_1 + d_2)^3}{d_1 d_2} \tag{8.1}$$

where ε is the in-plane strain, and d_1 and d_2 are the thicknesses of the layers. In general, tubes prepared with optical applications in mind may have diameters on the order of several micrometers, although tubes with diameters as small as tens of nanometers have been demonstrated [32]. The tube rolling generally takes place along the (100) crystal direction in this system because of the anisotropy of the Young's modulus in this kind of crystals [33]. Not only can the tube diameter be controlled but also its final surface geometry. This can be achieved by patterning the nanomembrane [34,35]. For instance, when patterned in the shape of high aspect ratio rectangles, the membrane will roll along its short side. For rectangles with a lower aspect ratio,

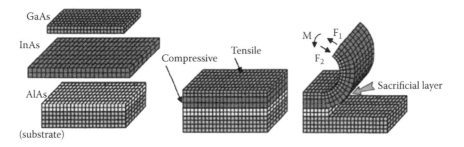

FIGURE 8.1 Illustration of the rolling mechanism for a strained InAs/GaAs bilayer membrane. (From Li, X., *J. Phys. D: Appl. Phys.*, vol. 41, p. 193001, 2008. With permission.)

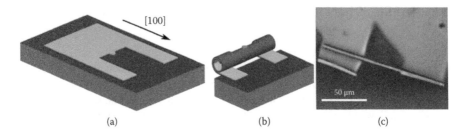

FIGURE 8.2 Illustration of the fabrication of free-standing semiconductor tube optical cavities. (a) Lithographically defined U-shaped mesa. (b) Rolled-up tube resulting from the sacrificial etching and release of the U-shaped mesa. (c) Optical microscopy image of a rolled-up InGaAsP tube.

the rolling occurs preferentially along the long side [34]. In addition, the starting rolling edge and resulting geometry can be altered by external means such as pinning the membrane with photoresist [24].

The semiconductor tube fabrication combines the top-down and bottom-up approaches. Starting from a strained nanomembrane, standard photolithography and etching top-down processes are used to define a starting edge for the rolling process. In crystalline materials, the associated mesas need to have the proper crystallographic alignment for controlled rolling to occur (for instance, along the aforementioned (100) axis in crystalline semiconductor III-V systems [34]). To realize free-standing tube cavities, a U-shaped mesa, illustrated in Figure 8.2a, is commonly employed. In this scheme, both the thickness of the free-standing tube and the vertical separation from the substrate can be entirely controlled by the shape of the mesa. The tubes will then self-assemble (in a bottom-up fashion) by means of a sacrificial etching process. In the case of InGaAs/GaAs nanomembranes, an AlAs layer is usually included and used as a sacrificial layer [17,35]. The sacrificial etching process will etch away this distinct layer and release the strained InGaAs/GaAs membrane. In other material systems, such as InP/InGaAsP [24], the availability of good selective etchants allows the substrate to play the role of the sacrificial layer. Figure 8.2b,c illustrates the resulting microtube after the etching of the U-shaped mesa. As can be appreciated in these figures, the edges of the mesa perpendicular to the rolling direction (the "rolling edges") will result in steps in the thickness of the microtube wall, which play an important role in the emission characteristics of rolled-up semiconductor tube cavities, described in the following sections.

There have also been many reports on the fabrication of rolled-up tube structures using dielectric and metal membranes. For instance, tubes based on strained SiO_x/Si membranes, which show photoluminescence in the near-infrared range, have been demonstrated [25]. Polymer tubes fabricated using deposited or electrospun polymers have also been reported [29,36,37]. Single and bimetallic (Au, Ti) tubes have been fabricated from self-rolled polymer templates [38]. Combining different kinds of materials, hybrid rolled-up tubes have been reported such as InGaAs/GaAs/Au [39] and InGaAs/GaAs/Nb [40] metallized tubes, and InGaAs/GaAs/polymer hybrid organic-semiconductor tubes [41].

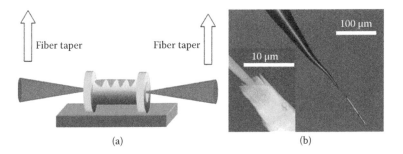

(a)   (b)

FIGURE 8.3 (a) Schematic illustration of the transfer method with the use of abruptly tapered optical fibers. (b) Optical microscopy image of a semiconductor tube on the tip of an abrupt fiber taper. Inset: Scanning electron microscopy image of the transition between the glass fiber and the tube. (Tian, Z., et al., *IEEE Photon. Technol. Lett.*, vol. 22, p. 311, 2010. With permission.)

8.3 TRANSFER OF ROLLED-UP TUBE STRUCTURES

To take advantage of the already mature silicon complementary metal-oxide semiconductor [CMOS] technology and simultaneously benefit from the III-V semiconductor characteristics such as direct bandgap and high carrier mobility, it is necessary to use a proper technique to transfer the active III-V devices onto silicon substrates. Although special techniques such as wafer bonding [42,43] and dry printing [44–46] have been developed, it is not possible to use them for rolled-up microtubes as they tend to break during the transfer process. In this regard, certain methods have been demonstrated to effectively transfer the microtubes. In the recently reported substrate-on-substrate transfer process [47], the host substrate (GaAs wafer), with the presence of free-standing InGaAs/GaAs tubes, is pressed on top of a silicon wafer in the presence of a solvent. By removing the GaAs substrate, the tubes preferentially stay on the Si substrate due to the gravitational force induced by the solvent around the tube structure. Thanks to surface tension forces, the tubes are subsequently attached to the Si substrate. Alternatively, rolled-up tube structures can also be transferred on foreign substrates by first dispersing them in a solvent solution, which is then drop cast on the substrate [32]. A unique fiber-taper assisted transfer process has been developed to achieve precise control over the transfer process [20]. In this approach, abruptly tapered optical fibers are inserted at one or both ends of the tube structure and are used as a handle to pick up the tube from its host substrate. The transfer process is schematically illustrated in Figure 8.3. Subsequently, the tube can be transferred to a foreign substrate with precise positioning, compared to other transfer processes. Using this transfer technique, direct integration of rolled-up optical tube cavities with silicon-on-insulator waveguides has been demonstrated [48].

8.4 OPTICAL CHARACTERISTICS

With the formation of rolled-up semiconductor tubes, improvement in the optical properties, such as enhancement in the photoluminescence intensity of the quantum wells or dots embedded in the tube structures, has been commonly observed [49–51].

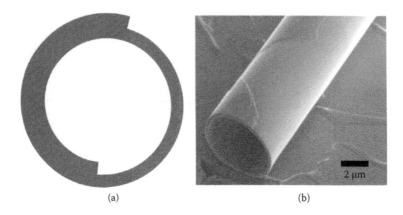

(a)   (b)

FIGURE 8.4 (a) Illustration of the spiral geometry of a rolled-up tube, showing the presence of inner and outer notches. (b) Scanning electron microscopy image of the cross-section of an InGaAs/GaAs tube. (From Li, F. and Mi, Z., *Opt. Express*, vol. 17, p. 19933, 2009. With permission.)

Coupling between the photoluminescence emission and the cavity resonant modes can lead to regularly spaced emission peaks for these cavities. The optical modes in a rolled-up tube cavity are directly related to the azimuthal, axial, and radial confinements [17,52]. Consequently, the emission characteristics can be varied by controlling the tube diameter, surface geometry, and wall thickness. Compared to conventional optical cavities with a circular cross-section, such as microrings [10], microdisks [9], and microspheres [53], rolled-up tube cavities exhibit spiral symmetry due to the presence of inside and outside notches, illustrated in Figure 8.4, which break the degeneracy for optical modes propagating in the clockwise and counterclockwise directions [19]. Additionally, photons are largely scattered from the inside notch, thereby leading to optical microcavities with directional emission [18]. It has also been observed that rolled-up semiconductor tubes with relatively thin walls can only support transverse electric (TE) modes, with the electric field polarized along the tube surface [52].

8.4.1 ELECTROMAGNETIC MODELING OF ROLLED-UP TUBE CAVITIES

The emission characteristics of rolled-up tube cavities can be modeled to better understand their optical properties. The most commonly used techniques are the finite-difference time-domain (FDTD) method [54] and a simplified planar dielectric waveguide model [16,19,21], as described next.

8.4.1.1 Finite-Difference Time-Domain (FDTD) Analysis

Fully vectorial numerical simulations can be applied to optical resonators to compute resonant mode frequencies and field distributions [55]. FDTD is one of the most popular methods for this purpose due to its simplicity and ease of use, and it is an accurate method for studying electromagnetic problems including the simulation of optical cavities [56]. In this study, we have restricted our simulations to two dimensions where we compute the spectra and fields for the cross-section of

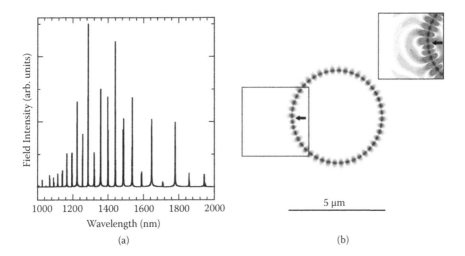

(a)　　　　　　　　　　　　　　　　　(b)

FIGURE 8.5 (a) Calculated resonance mode distribution of a tube cavity using a two-dimensional finite-difference time-domain algorithm. (b) Calculated electric field distribution, plotted using a linear color scale to emphasize the field confinement. The black arrow marks the position of the inner rolling edge. Inset: Detail of the marked square, using a modified brightness scale to emphasize the directional emission pattern.

a rolled-up tube. Although this simplified approach does not capture the axial behavior of the modes, as the model only considers in-plane wave vectors (i.e., $k_z = 0$), other features of the rolled-up geometry can be analyzed. For example, it allows identification of azimuthal modes and their corresponding mode numbers.

An example of a FDTD-based computation on a tube is shown in Figure 8.5. The mode spectrum shown in Figure 8.5a was calculated by using a broadband dipole source. One of the strongest modes at the wavelengths of interest (near 1425 nm in this case) was selected, and a narrowband dipole source at that wavelength was used to excite it. After the buildup of the mode, a snapshot of the electric field intensity, shown in Figure 8.5b, was captured. Since the FDTD simulations compute the full field of the mode, it is possible to use it to study many features. For instance, we can analyze the directionality of the emission. Rescaling the computed field, it is seen (shown in the inset in Figure 8.5b) that the emission occurs preferentially from the inner rolling edge of the tube, rather than being isotropic as is the case for an ideal ring resonator [18,19]. The FDTD calculations also illustrate well the optical confinement in the microtube walls and further allow quantification of this confinement by computing Q-factors.

8.4.1.2 Equivalent Planar Waveguide Model

It is possible to model the full optical behavior of the tubes using an equivalent planar waveguide model [16,17]. This three-dimensional model requires much less resources than a fully vectorial simulation. In this model the tube is first "unrolled"; then the field of the resulting planar waveguide is derived. The resonant wavelengths of the azimuthal, axial, and radial modes are subsequently obtained by applying periodic boundary conditions. This resulting model is illustrated in Figure 8.6,

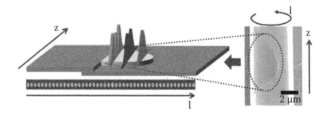

FIGURE 8.6 SEM image of a rolled-up microtube (right) and schematic of the equivalent planar dielectric waveguide model (left). The simulated azimuthal optical mode distribution for a simple ring resonator of the same diameter is also illustrated along the waveguide. (From Li, F., et al., *Opt. Lett.*, vol. 34, p. 2915, 2009. With permission.)

wherein the tube is converted into a waveguide with a step on its thickness. Also shown in Figure 8.6 is a parabolic-like surface corrugation. This surface geometry plays a critical role in the axial confinement of light. There are two distinguishable parts in the unrolled waveguide, including a thin part with a length L_{thin}, and a thick one that has a length L_{thick}, whose thickness difference is equal to the thickness of the source nanomembrane. These lengths are such that $L_{thin} + L_{thick} = 2\pi R$ (where R is the microtube radius). In order to simplify the calculation process, well-known solutions for slab waveguides [57] are used to obtain the effective indices corresponding to the "thin" and "thick" parts of the waveguide ($n_{thin}(z)$ and $n_{thick}(z)$, respectively), which can vary along the resonator long axis (z-axis). To further simplify the calculation, the waveguide is considered to have an average thickness, with an effective index $n_{avg}(z) = [L_{thin}n_{thin}(z) + L_{thick}n_{thick}(z)]/2\pi R$. If the tube wall is relatively thick, the equivalent waveguide may support more than one mode. These modes correspond to different radial modes in the rolled-up tube cavity. For thin-walled tubes, only the fundamental TE mode, with its electrical field parallel to the tube surface, may be supported by the waveguide [35]. In this case, the resonant modes can be derived by solving the scalar Helmholtz equation [58],

$$\frac{1}{n_{avg}^2(z)}\left(\frac{d^2E(l,z)}{dz^2} + \frac{d^2E(l,z)}{dl^2}\right) = k^2E(l,z) \qquad (8.2)$$

where $E(l,z)$ is the electric field and $k = 2\pi/\lambda_0$ is the vacuum wave vector. If the variation of $n_{avg}(z)$ along the z-axis is relatively small [16], separation of variables is valid and a solution of the form $E(l,z) = \varphi(z)\exp(i\beta l)$ can be postulated for the electric field. These steps lead to the "photonic quasi-Schrödinger" equation,

$$-\frac{d^2\phi}{dz^2} - n_{avg}^2(z)k^2\phi^2 = -\beta^2\phi(z) \qquad (8.3)$$

Equation (8.3) can be solved numerically for a general axial perturbation. An analytic solution can also be derived when the $n_{avg}(z)$ profile is parabolic [16]. The solutions for Equation (8.3) will be a set of dispersion relations for different transverse modes in the planar waveguide, wherein each mode corresponds to a different

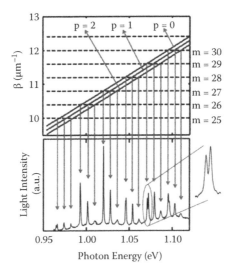

FIGURE 8.7 Room temperature photoluminescence spectrum (lower graph) and calculated spectral eigenmodes (upper graph) of a single-walled quantum-dot microtube ring resonator. Solid lines: calculated dispersion curves for the first three transverse optical modes ($p = 0, 1, 2$) using an equivalent planar dielectric waveguide model; dashed lines: resonance conditions for various azimuthal modes ($m = 25, 26, 27, 28, 29, 30$). The intersections between the solid and dashed lines correspond to the cavity eigenmodes. (From Li, F., et al., *Opt. Lett.*, vol. 34, p. 2915, 2009. With permission.)

axial mode, denoted by the axial mode number p, of the microtube. The available modes are then reduced to a discrete set of wavelengths by applying periodic boundary conditions on the waveguide,

$$\beta R = m \tag{8.4}$$

where m is an integer known as the azimuthal mode number. As an example, the computed dispersion relations for different transverse modes of a rolled-up InAs/GaAs tube with a diameter of ~5 μm and average wall thickness of ~50 nm is shown in the upper panel of Figure 8.7. The derived cavity eigenmodes are in excellent agreement with the experimental results, illustrated in the lower inset of Figure 8.7 [17].

8.4.1.3 Control of the Optical Resonance Modes

Compared to other optical microresonators, emission characteristics of rolled-up tube cavities can be readily tailored during the device fabrication process. First, the layered structures that form the tubes can be engineered to determine the tube diameters (see Equation 8.1). The diameter, as is the case with regular ring resonators, will largely determine the free spectral range (the spectral separation between consecutive azimuthal modes) according to

$$\text{FSR} \approx \frac{\lambda^2}{\pi n D} \tag{8.5}$$

where λ is the resonant wavelength, n is the effective refractive index, and D is the microtube diameter. Thus the basic resonance mode characteristics can be determined, to a large extent, during the material's growth/synthesis process. The other important parameter that can be varied is the wall thickness, which is specified using lithography and is related to the number of windings that will form the rolled-up tube. The wall thickness affects the radial mode properties as well as the polarization of the confined photons.

Additionally, the axial mode profile of the tube cavity can be tailored. From Equation (8.3) it is seen that the profile of the averaged refractive index along the tube axis will change the resulting optical modes. The axial index profile can be controllably varied by defining a pattern around the inner or outer edge of the mesa during the device fabrication process [16,17]. Properly designed patterns will induce confinement along the axial direction that can lead to three-dimensionally confined optical modes. Careful tuning of the pattern offers detailed control of the separation between these axial modes. As an example, the emission spectra for two InGaAsP/InAs quantum-dot tubes simultaneously fabricated from the same starting bilayer are shown in Figure 8.8.

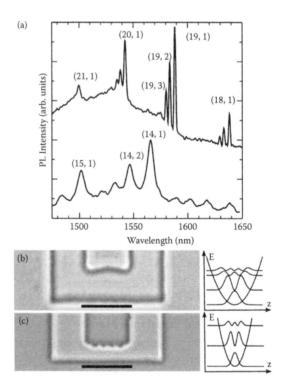

FIGURE 8.8 (a) Photoluminescence spectrum for rolled-up microtubes with an average thickness of two (top) and one (bottom) windings and parabolic-like surface geometry. The mode numbers were derived using the equivalent planar waveguide model. (b) Optical microscopy image of the lithographically designed mesa for the two windings microtube (left) and a schematic of the electric field of the first few axial modes (right). (c) Same as (b), but for the one winding microtube.

The first tube (top spectrum in Figure 8.8a) has an average thickness of two windings and a shallow parabolic-like axial profile (shown in Figure 8.8b). Its spectrum shows sharp, closely spaced peaks. The second tube (bottom spectrum in Figure 8.8a) has a thinner wall, with an average thickness of approximately one winding and a sharp parabolic-like axial profile (displayed in Figure 8.8c). Correspondingly, the peaks are broader and show a larger spacing. The right-side panels in Figure 8.8b,c illustrate the electric field of the first few axial modes in each case.

8.4.2 Transmission Properties

The transmission properties of rolled-up tube optical cavities have also been studied. For instance, it has been reported that rolled-up InGaAs/GaAs tubes can couple to silicon-on-insulator (SOI) waveguides [48]. In these experiments, a single tube was transferred onto a SOI waveguide, shown in Figure 8.9a.

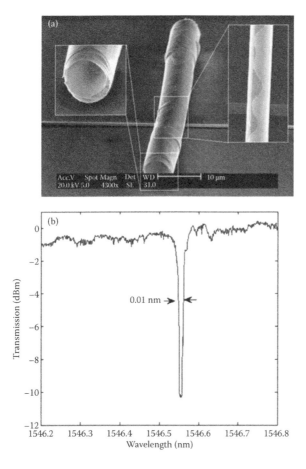

FIGURE 8.9 (a) Scanning electron microscopy image of an InGaAs/GaAs tube transferred on top of an SOI waveguide. (b) Detailed transmission spectrum showing a very sharp resonant dip. (From Tian, Z., et al., *Opt. Express*, vol. 19, p. 12164, 2011. With permission.)

The relative position of the tube and the waveguide was such that the field propagating along the waveguide could couple into resonant modes of the rolled-up tube cavity. The wavelength of a tunable laser diode coupled to the waveguide was scanned. If the laser wavelength was matched to that of a resonant mode, light from the waveguide was coupled into the tube, appearing as a dip in the measured transmission spectrum, illustrated in Figure 8.9b. The measured minimum line-width is ~0.01 nm, corresponding to a Q factor of 1.5×10^5. Transmission spectra have also been measured using adiabatically tapered optical fibers instead of SOI waveguides. Tapered optical fibers allow for control of the excitation polarization, as opposed to integrated waveguides, which usually support only one propagating polarization. Using this method, selective excitation of modes with different polarizations (TE and TM) was reported [59] in InGaAs/GaAs rolled-up tubes with relatively thick (~300 nm) walls. This selective mode excitation was further used as part of a light–light modulating system based on the absorption of pump light in the tube [59].

8.5 ROLLED-UP TUBE LASERS

Semiconductor rolled-up tubes offer several advantages for laser applications. As described earlier, quantum wells and quantum dots can be embedded in rolled-up semiconductor tubes to serve as the gain medium. In addition, rolled-up tube lasers can exhibit controlled polarization and emission direction. Another advantage is the large overlap between the confined optical field and the gain medium, which can lead to ultralow lasing threshold.

8.5.1 GaAs-Based Tube Lasers

The first demonstrations of tube lasers have been realized in the InGaAs/GaAs material system. Initial experiments, reported by Li and Mi [21], demonstrated a continuous wave, optically pumped tube laser operating at room temperature, wherein self-organized InGaAs quantum dots were incorporated as the active medium. The emission spectra of the device (both below and above the lasing threshold) are shown in Figure 8.10a. Mode numbers were calculated using the equivalent planar waveguide model (16,17). The above-threshold spectrum shows multimode emission in the 1200 to 1250 nm wavelength range. Figure 8.10b shows the light–light curve for the lasing mode (m = 37), and its corresponding linewidth. The lasing threshold was estimated to be ~4 μW. Considering the peak as a composite of two overlapping mode peaks due to the spiral asymmetry of the tube cavity [19], the minimum lasing linewidth was estimated to be between 0.2 and 0.3 nm. The linewidth displayed an increase at higher excitation powers, which was attributed to heating of the tube [21]. Time-resolved studies of an InAlGaAs/AlGaAs tube laser have also been reported [22]. This study used GaAs quantum wells as the gain medium, and the tube was optically pumped using a pulsed laser at a temperature of ~10 K. The device showed single-mode lasing, with a threshold between 260 and 595 μW and an emission wavelength near 800 nm. Interestingly, even with pulsed pumping at low temperatures, heating

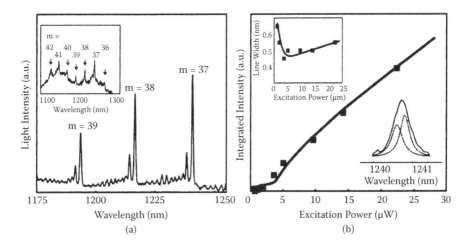

FIGURE 8.10 Emission spectrum of InGaAs/GaAs quantum dot microtube lasers measured at an absorbed pump power of ~23 μW (above threshold). The emission spectrum measured at an absorbed pump power of ~ 3 μW (below threshold) is shown in the inset. (b) The integrated light intensity for lasing mode at 1240.7 nm versus absorbed pump power at room temperature. Variation of the linewidth of the mode versus absorbed pump power is shown in the upper inset. A detailed view of the optical resonance mode at ~1240.7 nm above threshold and the fit with two Lorentzian curves are shown in the lower inset. (From Li, F. and Mi, Z., *Opt. Express*, vol. 17, p. 19933, 2009. With permission.)

in the tube is also detected at high excitation powers, leading to a red shift in the emission wavelength.

8.5.2 INP-BASED TUBE LASERS

More recently, an InGaAsP tube laser, with emission in the telecom S band, has been demonstrated [66]. This laser used InAs quantum dots as a gain medium, and was optically pumped with a continuous wave laser at liquid nitrogen temperature. The tube device has a wall thickness of ~100 nm and diameter of ~5 μm. The emission spectra measured at ~0.18 μW (below threshold) and ~5.8 μW (above threshold) are shown in Figure 8.11a. The mode numbers were derived using the equivalent planar waveguide model. Also shown in the figure, as a reference, is the photoluminescence spectrum measured from as-grown InAs/InGaAsP quantum-dot heterostructures. Figure 8.11b depicts the light–light curve for the mode labeled (22,1), showing a clear kink near the threshold pump power, from which a threshold power of ~1.26 μW was derived. The mode linewidth showed a clear decrease above threshold, shown in the inset of Figure 8.11b, signaling its increased temporal coherence. Finally, the spontaneous emission background was calculated from a 4 nm wide spectral region, shown as the square in Figure 8.11a. Variations of this background emission with pumping power are plotted in Figure 8.11b, which shows a much lower rate of growth for the spontaneous emission with pump power, compared to the lasing mode emission above threshold, which further indicates the achievement of lasing.

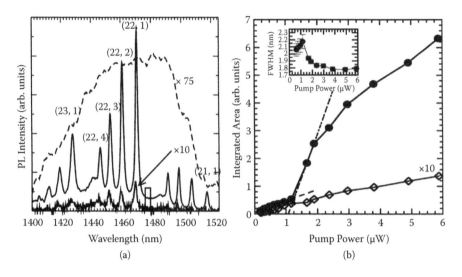

FIGURE 8.11 (a) Emission spectrum for an InAs/InGaAsP quantum dot microtube device measured at 82 K. The lower (weaker) spectrum corresponds to a small absorbed pump power of 180 nW (below threshold), which was multiplied by a factor of 10 for improved visibility. The upper (stronger) spectrum was measured at a high absorbed pump power of ~5.6 μW (above threshold). The dashed line shows the emission of the as-grown InAs/InGaAsP quantum-dot sample (multiplied by a factor of 75) as a reference. (b) Light–light curve (closed circles) for the (22,1) mode. The integrated spontaneous emission background calculated from the rectangular box in panel (a) is plotted, multiplied by a factor of 10, using open diamonds. Inset: Linewidth of the lasing mode as a function of the absorbed pump power. (From Bianucci, P., et. al., Self-organized InAs/InGaAsP quantum dot tube lasers, *Appl. Phys. Lett.* vol. 101, p. 031104, 2012 [66].)

8.6 EMERGING APPLICATIONS

8.6.1 OPTICAL COMMUNICATIONS

The achievement of lasing in rolled-up tubes, together with their small dimensions, tunability, excellent lasing characteristics, and direct transfer on foreign substrates render them ideal laser sources in optical communications. For example, InAs/InGaAsP tube devices transferred on a Si platform with emission in the S (1460–1530 nm) and C (1530–1565 nm) optical telecommunication bands would be suitable as light sources in chip-level optical communications. Additionally, rolled-up tubes could also function as modulators and add-drop filters [59].

8.6.2 MICROFLUIDIC BIOSENSORS

The fabrication flexibility of rolled-up tubes makes it possible to create optical resonators with subwavelength wall thicknesses, leading to a highly enhanced evanescent field. As a consequence, the resonant modes in the device are ultrasensitive to optical perturbations outside its walls (to a distance determined by the exponential decay of this evanescent field, usually on the order of several hundred nanometers) [23], thereby promising optical sensing devices capable of detecting minute changes in the refractive index of their environment. Huang et al. [23] demonstrated this concept by measuring the changes in the emission spectrum of SiO/SiO_2 microtubes as a function of their environment. Measurements taken in air, ethanol, water, and a mixture of the last two showed a shift of the resonant mode wavelengths to the red as well as a broadening of the mode peaks when the medium refractive index was increased. This particular demonstration reached a sensitivity of 425 nm/refractive index units (RIU), with a detection limit of 10^{-4} RIU.

The hollow geometry of rolled-up tubes immediately suggests using them in tandem with microfluidics, wherein the liquid to be sensed can flow through the channel defined by the tube hollow interior. Routing of fluid and coupling to a microfluidic system have already been demonstrated [60,61], showing the possibility for realizing an integrated rolled-up-tube-based microfluidic sensing system.

8.6.3 BEYOND OPTICS

Beyond their optical properties, rolled-up tubes have been shown to have many other promising applications. For instance, using a set of strained layers in a conductor–dielectric–conductor configuration, it is possible to fabricate rolled-up supercapacitors. As an example, a RuO_2 rolled-up micro-supercapacitor has been realized, which exhibited a 7 μm diameter and 1.0 μF capacitance [62]. These results have shown a reduction in size of two orders of magnitude compared to existing commercial capacitors with a similar capacitance. Combining large numbers of high-capacitance tubes into a single device could result in compact supercapacitors with an extraordinary ability to hold on to electrical charge.

Another very interesting development is the demonstration of catalytic microjet engines [63]. For this purpose, the tube structure is designed in such a way that it is slightly tapered (looking more like a hollow micronozzle) with its inside surface including a catalytic film (such as platinum). When placed in a hydrogen peroxide solution, the catalytic action of the platinum decomposes the peroxide in water and gaseous oxygen, which forms a microbubble. This bubble then moves toward the wide end of the tube, where it exits. The mechanical reaction to the exiting gas pushes the microtube forward in the fluid, causing fresh peroxide to enter and maintaining the motion. The addition of magnetic materials to the tube allows steering of the tube motion by applying external magnetic fields [64].

Finally, rolled-up tube structures can also be used as scaffolding for the growth of cells [65], where a rolled-up tube fabricated with a biocompatible material has been used to guide the growth of yeast cells and study their behavior in confined spaces.

8.7 CONCLUSION

In this chapter we briefly reviewed the current progress in semiconductor rolled-up tubes, with an emphasis on their optical properties and applications. We discussed their fabrication, which involves top-down patterning and bottom-up self-assembly. We introduced techniques to model their optical properties, both semianalytically (the equivalent planar waveguide model) and numerically (finite-difference time-domain method) and discussed their resonant optical properties. From this discussion, it has become clear that rolled-up tubes offer a large degree of flexibility in tuning their optical spectral properties. The ability to transfer the tubes onto foreign substrates and further integrate with standard SOI waveguides showcases their potential as integrated optical devices. Ultralow threshold rolled-up tube lasers have also been demonstrated, which can potentially emerge as an integrated laser source for on-chip optical communications. Finally, with the wide range of applications that rolled-up tubes have already shown, it becomes obvious that at this point in time it is appropriate to talk about them as a technology or platform rather than just a device concept.

REFERENCES

1. Chang, R. K., and Campillo, A. J., eds. *Optical processes in microcavities.* Singapore: World Scientific, 1996.
2. Purcell, E. M. Spontaneous emission probabilities at radio frequencies. *Phys. Rev.,* vol. 69, p. 681, 1946.
3. Koch, S. W., Jahnke, F., and Chow, W. W. Physics of semiconductor microcavity lasers. *Semicond. Sci. Technol.,* vol. 10, p. 739, 1995.
4. Peter, E., et al. Exciton-photon strong coupling regime for a single quantum dot embedded in a microcavity. *Phys. Rev. Lett.,* vol. 95, p. 067401, 2005.
5. Bhattacharya, P., and Mi, Z. Quantum dot optoelectronic devices. *Proc. IEEE,* vol. 95, p. 1723, 2007.
6. Poole, P. J., et al. Chemical beam epitaxy growth of self-assembled InAs/InP quantum dots. *J. Vac. Sci. Technol. B*, vol. 19, p. 1467, 2001.
7. Coleman, J. J., Beernink, K. J., and Givens, M. E. Threshold current density in strained layer InxGa(1-x)As-GaAs quantum well heterostructure lasers. *IEEE J. Quantum Electron.,* vol. 28, p. 1983, 1992.
8. Ellis, B., et al. Ultralow-threshold electrically pumped quantum-dot photonic-crystal nanocavity laser. *Nature Photon.,* vol. 5, p. 297, 2011.
9. Van Campenhout, J., et al. Electrically pumped InP-based microdisk lasers integrated with a nanophotonic silicon-on-insulator waveguide circuit. *Opt. Express*, vol. 15, p. 6744, 2007.
10. Liang, D., et al. Electrically pumped compact hybrid silicon microring lasers for optical interconnects. *Opt. Express*, vol. 17, p. 20355, 2009.
11. Reitzenstein, S., et al. Lasing in high-Q quantum-dot micropillar cavities. *Appl. Phys. Lett.,* vol. 89, p. 051107, 2006.
12. Loncar, M., et al. Low-threshold photonic crystal laser. *Appl. Phys. Lett.,* vol. 81, p. 2680, 2002.
13. Chen, R., et al. Nanolasers grown on silicon. *Nature Photon.,* vol. 5, p. 175, 2011.
14. Yu, K., Lakhani, A., and Wu, M. C. Subwavelength metal-optic semiconductor nanopatch lasers. *Opt. Express*, vol. 18, p. 8790, 2010.
15. Li, X. Strain induced semiconductor nanotubes: From formation process to device applications. *J. Phys. D: Appl. Phys.,* vol. 41, p. 193001, 2008.
16. Strelow, C., et al. Optical microcavities formed by semiconductors using a bottlelike geometry. *Phys. Rev. Lett.,* vol. 101, p. 127403, 2008.
17. Li, F., Mi, Z., and Vicknesh, S. Coherent emission from ultrathin-walled spiral InGaAs/InAs quantum dot microtubes. *Opt. Lett.,* vol. 34, p. 2915, 2009.
18. Strelow, C., et al. Spatial emission characteristics of a semiconductor microtube ring resonator. *Physica E*, vol. 40, p. 1836, 2008.

19. Hosoda, M., and Shigaki, T. Degeneracy breaking of optical resonance modes in rolled-up spiral microtubes. *Appl. Phys. Lett.,* vol. 90, p. 181107, 2007.
20. Tian, Z., et al. Controlled transfer of single rolled-up InGaAs-GaAs quantum-dot microtube ring resonators using optical fiber abrupt tapers. *IEEE Photon. Technol. Lett.,* vol. 22, p. 311, 2010.
21. Li, F., and Mi, Z. Optically pumped rolled-up InGaAs/GaAs quantum dot microtube lasers. *Opt. Express*, vol. 17, p. 19933, 2009.
22. Strelow, C., et al. Time-resolved studies of a rolled-up semiconductor microtube laser. *Appl. Phys. Lett.*, vol. 95, p. 221115, 2009.
23. Huang, G., et al. Rolled-up optical microcavitites with subwavelength wall thicknesses for enhanced liquid sensing applications. *ACS Nano*, vol. 4, p. 3123, 2012.
24. Mi, Z., et al. Self-organized InAs quantum dot tube lasers and integrated optoelectronics in Si. *Proc. SPIE,* vol. 7943, p. 79431C, 2011.
25. Songmuang, R., et al. SiOx/Si radial superlattices and microtube optical ring resonators. *Appl. Phys. Lett.,* vol. 90, p. 091905, 2007.
26. Vorob'ev, P., et al. SiGe/Si microtubes fabricated on a silicon-on-insulator substrate. *J. Phys. D: Appl. Phys.*, vol. 36, p. L67, 2003.
27. Mei, Y., et al. Fabrication, self-assembly, and properties of ultrathin AlN/GaN porous crystalline nanomembranes: Tubes, spirals and curved sheets. *ACS Nano*, vol. 3, p. 1663, 2009.
28. Muller, C., et al. Tuning magnetic properties by roll-up of Au/Co/Au films into micro-tubes. *Appl. Phys. Lett.*, vol. 94, p. 102510, 2009.
29. Dror, Y., et al. One-step production of polymeric microtubes by co-electrospinning. *Small*, vol. 3, p. 1064, 2007.
30. Prinz, V. Y., et al. Free-standing and overgrown InGaAs/GaAs nanotubes, nanohelices and their arrays. *Physica E*, vol. 6, p. 828, 2000.
31. Deneke, C., et al. Diameter scalability of rolled-up In(Ga)As/GaAs microtubes. *J. Semicond. Sci. Tech.*, vol. 17, p. 1278, 2002.
32. Chun, I. S., and Li, X. Controlled assembly and dispersion of strain-induced InGaAs/GaAs nanotubes. *IEEE Trans. Nanotech.*, vol. 7, p. 493, 2008.
33. Cottam, R. I., and Saunders, G. A. The elastic constants of GaAs from 2 K to 320 K. *J. Phys. C: Solid State Phys.*, vol. 6, p. 2015, 1973.
34. Chun, I. S., et al. Geometry effect on the strain-induced self-rolling of semiconductor membranes. *Nano Lett.,* vol. 10, p. 3927, 2010.
35. Kipp, T., et al. Optical modes in semiconductor microtube ring resonators. *Phys. Rev. Lett.*, vol. 96, p. 077403, 2006.
36. Luchnikov, V., et al. Focused-ion-beam-assisted fabrication of polymer rolled-up micro-tubes. *J. Micromech. Microeng.*, vol. 16, p. 1602, 2006.
37. Zhang, H., et al. Hybrid microtubes of polyoxometalate and fluorescence dye with tunable photoluminescence. *Chem. Commun.*, vol. 48, p. 4462, 2012.
38. Kumar, K., et al. Fabrication of metallic microtubes using self-rolled polymer tubes as templates. *Langmuir*, vol. 25, p. 7667, 2009.
39. Monti, G., et al. Metallic rings in a self-rolled micro-tube for magnetic field mapping applications. *2010 European Microwave Conference* (EuMC), p. 1385.
40. Thurmer, D. J., Deneke, C., and Schmidt, O. G. In situ monitoring of the complex rolling behaviour of InGaAs/GaAs/nb hybrid microtubes. *J. Phys. D: Appl. Phys.,* vol. 41, p. 205419, 2008.
41. Giordano, C., et al. Hybrid polymer/semiconductor microtubes: A new fabrication approach. *Microelectron. Engin.*, vol. 85, p. 1170, 2007.
42. Tanabe, K., et al. Room temperature continuous wave operation of InAs/GaAs quantum dot photonic crystal nanocavity laser on silicon substrate. *Opt. Express*, vol. 17, p. 7036, 2007.

43. Tong, Q. Y., et al. A "smarter-cut" approach to low-temperature silicon layer transfer. *Appl. Phys. Lett.*, vol. 72, p. 49, 1998.

44. Menard, E., et al. A printable form of silicon for high performance thin film transistors on plastic substrates. *Appl. Phys. Lett.*, vol. 84, p. 5398, 2004.

45. Kim, D. H., et al. Stretchable and foldable silicon integrated circuits. *Science*, vol. 320, p. 507, 2008.

46. Yuan, H. C., et al. Flexible photodetectors on plastic substrates by use of printing transferred single crystal germanium membranes. *Appl. Phys. Lett.*, vol. 94, p. 013102, 2009.

47. Vicknesh, S., Li, F., and Mi, Z. Optical microcavities on Si formed by self-assembled InGaAs/GaAs quantum dot microtubes. *Appl. Phys. Lett.*, vol. 94, p. 081101, 2009.

48. Tian, Z., et al. Single rolled-up InGaAs/GaAs quantum dot microtubes integrated with silicon-on-insulator waveguides. *Opt. Express*, vol. 19, p. 12164, 2011.

49. Chun, I. S., et al. Tuning the photoluminescence characteristics with curvature for rolled-up GaAs quantum well microtubes. *Appl. Phys. Lett.*, vol. 96, p. 251106, 2010.

50. Hosoda, M., et al. Quantum-well microtube constructed from a freestanding thin quantum-well layer. *Appl. Phys. Lett.*, vol. 83, p. 1017, 2003.

51. Mendach, S., et al. Light emission and waveguiding of quantum dots in a tube. *Appl. Phys. Lett.*, vol. 88, p. 111120, 2006.

52. Strelow, C., et al. Three dimensionally confined optical modes in quantum-well microtube resonators. *Phys. Rev. B*, vol. 76, p. 045303, 2007.

53. Bianucci, P., et al. Whispering gallery modes in silicon nanocrystal coated microcavities. *Physica Status Solidi A*, vol. 206, p. 973, 2009.

54. Taflove, A., and Hagness, S. C. *Computational electrodynamics: The finite-difference time-domain method.* 3rd ed. Norwood, MA: Artech House, 2005.

55. Hagness, S. C., et al. FDTD microcavity simulations: Design and experimental realization of waveguide-coupled single-mode ring and whispering-gallery-mode disk resonators. *J. Lightwave Technol.*, vol. 15, p. 2154, 1997.

56. Rodriguez, J. R., et al. Whispering gallery modes in hollow cylindrical microcavities containing silicon nanocrystals. *Appl. Phys. Lett.*, vol. 92, p. 131119, 2008.

57. Snyder, A. W., and Love, J. D. *Optical waveguide theory.* London: Chapman & Hall, 1983.

58. Zhang, K., and Li, D. *Electromagnetic theory for microwaves and optoelectronics.* Berlin: Springer Verlag, 1998.

59. Tian, Z., et al. Selective polarization mode excitation in InGaAs/GaAs microtubes. *Opt. Lett.*, vol. 36, p. 3506, 2011.

60. Thurmer, D. J., et al. Process integration of microtubes for fluidic applications. *Appl. Phys. Lett.*, vol. 89, p. 223507, 2006.

61. Harazim, S. M., et al. Fabrication and applications of large arrays of multifunctional rolled-up SiO/SiO2 microtubes. *J. Mater. Chem.*, vol. 22, p. 2878, 2011.

62. Ji, H., Mei, Y., and Schmidt, O. G. Swiss roll nanocapacitors with controlled proton diffusion as redox micro-supercapacitors. *Chem. Commun.*, vol. 46, p. 3881, 2010.

63. Solovev, A. A., et al. Catalytic microtubular jet engines self-propelled by accumulated gas bubbles. *Small*, vol. 5, p. 1688, 2009.

64. Mei, Y., et al. Rolled-up nanotech on polymers from basic perception to self-propelled catalytic engines. *Chem. Soc. Rev.*, vol. 40, p. 2109, 2011.

65. Huang, G., et al. Rolled-up transparent microtubes as two-dimensionally confined culture scaffolds of individual yeast cells. *Lab Chip*, vol. 9, p. 263, 2008.

66. Bianucci, P., et. al. Self-organized InAs/InGaAsP quantum dot tube lasers. *Appl. Phys. Lett.* vol. 101, p. 031104, 2012.

Index

Printed and bound by CPI Group (UK) Ltd, Croydon, CR0 4YY

18/10/2024

01776256-0003